A Concise
Introduction to
Pure
Mathematics

A CONCISE INTRODUCTION TO PURE MATHEMATICS

MARTIN LIEBECK

CHAPMAN & HALL/CRC

Boca Raton London New York Washington, D.C.

Library of Congress Cataloging-in-Publication Data

Liebeck, M.W. (Martin W.), 1954-
   A concise introduction to pure mathematics / Martin Liebeck.
       p.   cm.
   Includes bibliographical references and indexes.
   ISBN 1-58488-193-3 (alk. paper)
   1. Mathematics. I. Title.

QA8.4.L47 2000
510—dc21   99-462376

© 2000 by Chapman & Hall/CRC

No claim to original U.S. Government works
International Standard Book Number 1-58488-193-3
Library of Congress Card Number 99-462376
Printed in the United States of America   2 3 4 5 6 7 8 9 0
Printed on acid-free paper

To Ann, Jonny and Matthew

Contents

Foreword

One of the great difficulties in teaching undergraduate mathematics at universities in the U.S. is the great gap between teaching students a set of algorithms (which is very often the bulk of what is learned in calculus) and convincing students of the power, beauty and fun of the basic concepts in mathematics.

Martin Liebeck's book, *A Concise Introduction to Pure Mathematics,* is one of the best I have seen of filling this gap. In addition to preparing students to go on in mathematics, it is also a wonderful choice for a student who will not necessarily go on in mathematics but wants a gentle but fascinating introduction into the culture of mathematics. Liebeck starts with the basics and introduces number systems. In particular he discusses the real numbers and complex numbers. He shows how these concepts are natural and important in solving natural problems. Various topics in analysis, geometry, number theory and combinatorics are introduced and are shown to be fun and beautiful. Starting from scratch, Liebeck develops interesting results which hopefully will intrigue the student and give encouragement to continue to study mathematics.

This book will give a student the understanding to go on in further courses in abstract algebra and analysis. The notion of a proof will no longer be foreign, but also mathematics will not be viewed as some abstract black box. At the very least, the student will have an appreciation of mathematics.

As usual, Liebeck's writing style is clear and easy to read. This is a book that could be read by a student on his or her own. There is a wide selection of problems ranging from routine to quite challenging.

While there is a difference in mathematical education between the U.K. and the U.S., this book will serve both groups of students extremely well.

Professor Robert Guralnick
Chair of Mathematics Department
University of Southern California
Los Angeles, California

Preface

I can well remember my first lecture as a mathematics undergraduate, back in the olden days. In it, we were told about something called "Russell's Paradox" — does the set consisting of all sets which do not belong to themselves belong to itself? — after which the lecturer gave us some rules called the "Axioms of Set Theory." I came out of the lecture somewhat baffled. The second lecture, in which we were informed that "a_n tends to l if, for every $\epsilon > 0$, there exists N such that for all $n \geq N$, $|a_n - l| < \epsilon$," was also a touch bewildering. In fact, the lecturers were pretty good, and bafflement and bewilderment eventually gave way to understanding, but nevertheless it was a fairly fierce introduction to the world of university pure mathematics.

Nowadays we university lecturers are less fierce animals, and mathematics courses tend to start with a much gentler introduction to pure mathematics. I have given such a course, entitled "Foundations of Analysis," at Imperial College for the past few years, to students in the first term of the first year of their degree (generally in Mathematics, or some joint degree including Mathematics). This book has grown out of that course. As well as being designed for use in a first university course, the book is also suitable for self-study. It could, for example, be read by students between school and university, or indeed by anybody with a reasonable background in school mathematics.

One of my aims is to provide a robust bridge between school and university mathematics. For a number of the topics covered, students may well have studied some of the basic material on this topic at school, but this book will generally take the topic much further, in a way that is interesting and stimulating (at least to me). For example, many will have come across the method of mathematical induction, and used it to solve some simple problems, like finding a formula for the sum $1 + 2 + 3 + \ldots + n$. But I doubt that many have seen how induction can be used to study solid objects whose faces all have straight edges, and to show that the only so-called regular solids are the famous five "Platonic solids" (the cube, tetrahedron, octahedron, icosahedron and dodecahedron), as is done in Chapter 9.

I generally enjoy things more if they come in bite-sized pieces, and accordingly I have divided the book into 19 short chapters. Each chapter is followed by a selection of exercises, ranging from routine calculations to some quite challenging problems.

When starting to study pure mathematics at university, students often have a refreshing sense of "beginning all over again." Basic structures like the real numbers, the integers, the rational numbers, and the complex numbers, must be defined and studied from scratch, and even simple and obvious-looking statements about them must be proved properly. For example, it probably seems obvious that if n is an integer (i.e., one of the whole numbers $0, 1, -1, 2, -2, 3, -3\ldots$), and n^2 is odd, then n must also be odd. But how can we write down a rigorous proof of this fact? Methods for writing down proofs of this and many other simple facts form one of the themes of Chapter 1, along with a basic introduction to the language of sets.

In Chapter 2, I define and begin to study three of the basic number systems referred to in the previous paragraph: the real numbers (which we start off by thinking of as points on an infinite straight line — the "real line"); the integers; and the rational numbers (which are the fractions $\frac{m}{n}$, where m and n are integers). It takes some effort to prove that there is at least one real number which is not rational — a so-called irrational number — but once this is done, one can see quite easily that there are many irrational numbers. Indeed, by the last chapter of the book, we shall understand the strange fact that, in a very precise sense, there are "more" irrational numbers than rational numbers (even though there are infinitely many of each).

In studying properties of the system of real numbers, it is sometimes helpful to have ways of thinking of them which are different from just "points on the real line." In Chapter 3, I introduce the familiar decimal notation for real numbers, which provides a visual way of writing them down, and can be useful in their general study. Chapters 4 and 5 carry on with our basic study of the real numbers.

In Chapter 6, I bring our last important number system into the action — the complex numbers. Students may well have met these before. We begin by introducing a symbol i, and defining $i^2 = -1$. A general complex number is a symbol of the form $a + bi$, where a and b are real numbers. We soon find that using complex numbers we can write down solutions of all quadratic equations, and then proceed to study other equations like $x^n = 1$. We also find some beautiful links between complex numbers and geometry in the plane. Chapter 7 takes the theory of equations much further. Solving quadratics is probably very familiar, but much less well known is the method for solving cubic equations given in this chapter. We then look at general polynomial equations (i.e., equations of the form $x^n + a_{n-1}x^{n-1} + \ldots + a_1 x + a_0 = 0$), and explore the amazing fact that every such equation has solutions which are complex numbers.

I have already mentioned the method of proof by mathematical induction, which is introduced in Chapter 8. This is a technique for proving statements involving a general positive integer n, such as "the sum of the first n odd positive integers is equal to n^2," or "the number of regions formed by n straight lines drawn in the plane, no two parallel and no three going through the same point, is equal to $\frac{1}{2}(n^2 + n + 2)$." The method of induction is actually rather more powerful than first meets the eye, and Chapter 9 is devoted to the proof by induction of an elegant result, known as Euler's formula, about the relationship between the numbers of corners, edges and faces of a solid object, whose faces all have straight edges. Euler's formula has all sorts of uses. For example, if you want to make a football by sewing together hexagonal and pentagonal pieces of leather, in such a way that each corner lies on three edges, then the formula implies that you will need exactly 12 pentagonal pieces, no more and no less. I could not resist going further in this chapter and showing how to use Euler's formula to study the famous Platonic solids mentioned earlier.

Chapter 10 is rather important. With our somewhat naive understanding of the real numbers up to here, as points on the real line, or decimals, it is difficult to see how to prove even such basic properties as the fact that every positive real number has a square root. In this chapter I introduce machinery for proving this and many other results. This machinery is based on a fundamental fact about the real numbers, which concerns upper and lower bounds, and can often seem somewhat mysterious at first sight. I endeavour to dispel some of this mystery by showing how the fundamental fact can be used to solve a number of natural problems, such as the square root problem I mentioned before. The chapter is intended as an appetizer for students' first course in a subject known as "Analysis," which is the study of the real numbers and functions defined on them.

Chapters 11 through 14 are all about possibly the most fascinating number system of all: the integers. Students will know what a prime number is — an integer greater than 1 which is only divisible by 1 and itself — and are quite likely aware of the fact that every integer greater than 1 is equal to a product of prime numbers, although this fact requires a careful proof. Much more subtle is the fact that such a prime factorization is unique — in other words, given an integer greater than 1, we can express it as a product of prime numbers in only one way. "Big deal! So what?", I hear you say. Well, yes, it is a big deal (so big that this result has acquired the grandiose title of "The Fundamental Theorem of Arithmetic"), and after proving it I try to show its significance by using it in the study of a number of problems; for instance, apart from 1 and 0, are there any squares which differ from a cube by just 1?

Chapter 15 is about methods of counting things. For example, suppose I have given the same lecture course for the last 16 years, and tell 3 jokes each year. I never tell the same set of 3 jokes twice. At least how many jokes do I know? To solve this and other important counting problems, we introduce binomial coefficients, which leads us into the Binomial and Multinomial Theorems.

After a little formal theory of sets and relations in Chapters 16 and 17, I introduce functions in Chapter 18. The book concludes with Chapter 19, in which I address some fascinating questions about infinite sets. When can we say that two infinite sets have the same "size"? Can we ever say that one infinite set has bigger "size" than another? These questions are answered in a precise and rigorous way, and some of the answers may appear strange at first sight; for example, the set of all integers and the set of all rational numbers have the same size, but the set of all real numbers has greater size than these. Chapter 19 closes with a beautifully subtle result which tells us that an infinite set always has smaller size than the set of all its subsets. The proof of this is based on the argument of Russell's Paradox; which brings me back to where I started....

Let me now offer some comments on designing a course based on the book. Crudely speaking, the book can be divided into four fairly independent sections, with the following "core" chapters:

Introduction to number systems and analysis : Chapters 1, 2, 3, 4, 5, 6, 8, 10

Theory of the integers : Chapters 11, 12, 14

Introduction to discrete mathematics : Chapters 15, 16

Functions, relations, and countability : Chapters 17, 18, 19

One could design a one- or two-semester course in a number of ways. For example, if the emphasis is to be on discrete mathematics, the core chapters to use from the first section would be 1, 2 and 8, and all the other sections would be core for such a course. On the other hand, if the course is intended to prepare students more for a future course in analysis, one should use all the chapters in the first and last sections. Overall, I would recommend incorporating all four sections into your course — it works well!

I would like to express my thanks to my father, Dr. Hans Liebeck, who read the entire manuscript and suggested many improvements, as well as saving me from a number of embarrassing errors; any errors that remain are of course his responsibility. And, finally, I thank generations of students at Imperial who have sat through my "Foundations of Analysis" course, and have helped me to hone the course into the sleek monster which has grown into this book.

Chapter 1

Sets and Proofs

This chapter contains some introductory notions concerning the language of sets, and methods for writing proofs of mathematical statements.

Sets

A *set* is simply a collection of objects, which are called the *elements* or *members* of the set. There are a number of ways of describing a set. Sometimes the most convenient way is to make a list of all the objects in the set, and put curly brackets around the list. Thus, for example,

$\{1, 3, 5\}$ is the set consisting of the objects 1, 3 and 5.

$\{$Fred, dog, 1.47$\}$ is the set consisting of the objects Fred, dog and 1.47.

$\{1, \{2\}\}$ is the set consisting of two objects, one being the number 1 and the other being the set $\{2\}$.

Often, however, this is not a convenient way to describe our set. For example, the set consisting of all the people who live in Denmark is for most purposes best described by precisely this phrase (i.e., "the set of all people who live in Denmark"); it is unlikely to be useful to describe this set in list form $\{$Sven, Inge, Jesper, ...$\}$. As another example, the set of all real numbers whose square is less than 2 is neatly described by the notation

$$\{x \mid x \text{ a real number}, x^2 < 2\} \, .$$

(This is to be read: "the set of all x such that x is a real number and $x^2 < 2$." The symbol "$\mid$" is the "such that" part of the phrase.) Likewise,

$$\{x \mid x \text{ a real number}, x^2 - 2x + 1 = 0\}$$

denotes the set consisting of all real numbers x such that $x^2 - 2x + 1 = 0$.

As a convention, we define the *empty set* to be the set consisting of no objects at all, and denote the empty set by the symbol $\emptyset$.

If S is a set, and s is an element of S (i.e., an object that belongs to S), we write

$$s \in S$$

and say s *belongs to S*. If some other object t does not belong to S, we write

$$t \notin S.$$

For example,

$1 \in \{1, 3, 5\}$ but $2 \notin \{1, 3, 5\}$,

if $S = \{x \mid x$ a real number, $0 \le x \le 1\}$, then $1 \in S$ but Fred $\notin S$,

$\{2\} \in \{1, \{2\}\}$ but $2 \notin \{1, \{2\}\}$,

$1 \notin \emptyset$.

Two sets are defined to be equal when they consist of exactly the same elements; for example,

$\{1, 3, 5\} = \{3, 5, 1\} = \{1, 5, 1, 3\}$,

$\{x \mid x$ a real number, $x^2 - 2x + 1 = 0\} = \{1\}$.

We say a set T is a *subset* of a set S if every element of T also belongs to S (i.e., T consists of some of the elements of S). We write $T \subseteq S$ if T is a subset of S, and $T \not\subseteq S$ if not. For example, if $S = \{1, \{2\}, \text{cat}\}$, then

$$\{\text{cat}\} \subseteq S, \ \{\{2\}\} \subseteq S, \ \{2\} \not\subseteq S.$$

As another example, the subsets of $\{1, 2\}$ are

$$\{1, 2\}, \{1\}, \{2\}, \emptyset.$$

(By convention, $\emptyset$ is a subset of every set.)

This is all we shall need about sets for the time being. This topic will be covered somewhat more formally later in Chapter 16.

Proofs

Consider the following mathematical statements:

(1) The square of an odd integer is odd. (By an *integer* we mean a whole number, i.e., one of the numbers $\ldots, -2, -1, 0, 1, 2, \ldots$.)

(2) No real number has square equal to -1.

(3) Every positive integer is equal to the sum of two integer squares. (The integer squares are 0, 1, 4, 9, 16, 25, and so on.)

Each of these statements is either true or false. Probably you have quickly formed an opinion on the truth or falsity of each, and regard this as "obvious" in some sense. Nevertheless, to be totally convincing, you must provide clear, logical proofs to justify your opinions.

To clarify what constitutes a proof, we need to introduce a little notation. If P and Q are statements, we write

$$P \Rightarrow Q$$

to mean that statement P implies statement Q. For example,

$x = 2 \Rightarrow x^2 < 6,$

it is raining $\Rightarrow$ the sky is cloudy.

Other ways of saying $P \Rightarrow Q$ are:

if P then Q (e.g., if $x = 2$ then $x^2 < 6$);

Q if P (e.g., the sky is cloudy if it is raining);

P only if Q (e.g., $x = 2$ only if $x^2 < 6$; it rains only if the sky is cloudy).

Notice that $P \Rightarrow Q$ does *not* mean that also $Q \Rightarrow P$; for example, $x^2 < 6 \not\Rightarrow x = 2$ (where $\not\Rightarrow$ means "does not imply"). However, for some statements P, Q, it *is* the case that both $P \Rightarrow Q$ and $Q \Rightarrow P$; in such a case we write $P \Leftrightarrow Q$, and say "P if and only if Q." For example,

$x = 2 \Leftrightarrow x^3 = 8,$

you are my wife if and only if I am your husband.

The *negation* of a statement P is the opposite statement, "not P," written as $\bar{P}$. Notice that if $P \Rightarrow Q$ then also $\bar{Q} \Rightarrow \bar{P}$ (since if $\bar{Q}$ is true then P cannot be true, as $P \Rightarrow Q$).

For example, if P is the statement $x = 2$ and Q the statement $x^2 < 6$, then $P \Rightarrow Q$ says "$x = 2 \Rightarrow x^2 < 6$," while $\bar{Q} \Rightarrow \bar{P}$ says "$x^2 \geq 6 \Rightarrow x \neq 2$." Likewise, for the other example above we have "the sky is not cloudy $\Rightarrow$ it is not raining."

Perhaps laboring the obvious, let us now make a list of the deductions that can be made from the implication "it is raining $\Rightarrow$ the sky is cloudy," given various assumptions:

Assumption	Deduction
it is raining	sky is cloudy
it is not raining	no deduction possible
sky is cloudy	no deduction possible
sky is not cloudy	it is not raining

Now let us put together some examples of proofs. In general, a proof will consist of a series of implications, proceeding from given assumptions, until the desired conclusion is reached. As we shall see, the logic behind a proof can take several different forms.

Example 1.1
Suppose we are given the following three facts:
    (a) I will be admitted to Greatmath University only if I am clever.
    (b) If I am clever then I do not have to work hard.
    (c) I have to work hard.
What can be deduced?

Answer Write G for the statement "I will be admitted to Greatmath University," C for the statement "I am clever," and W for the statement "I have to work hard." Then (a) says $G \Rightarrow C$, and (b) says $C \Rightarrow \bar{W}$. Hence,

$$W \Rightarrow \bar{C} \quad \text{and} \quad \bar{C} \Rightarrow \bar{G}.$$

Since W is true by (c), we deduce that $\bar{G}$ is true, i.e., I will not be admitted to Greatmath University (thank goodness).

Example 1.2
In this example we prove statement (1) above: the square of an odd integer is odd.

PROOF   Let n be an odd integer. Then n is 1 more than an even integer, so $n = 1 + 2m$ for some integer m. Therefore, $n^2 = (1+2m)^2 = 1+4m+4m^2 = 1 + 4(m + m^2)$. This is 1 more than $4(m + m^2)$, an even number, hence n^2 is odd. ■

Formally, we could have written this proof as the following series of implications:

n odd $\Rightarrow n = 1 + 2m \Rightarrow n^2 = 1 + 4(m + m^2) \Rightarrow n^2$ odd.

However, this is evidently somewhat terse, and such an approach with more complicated proofs quickly leads to unreadable mathematics; so, as in the above proof, we insert words of English to make the proof readable, including words like "hence," "therefore," "then" and so on, to take the place of implication symbols.

Note The above proof shows rather more than just the oddness of n^2: it shows that the square of an odd number is always 1 more than a multiple of 4, i.e., is of the form $1 + 4k$ for some integer k.

The proofs given for Examples 1.1 and 1.2 could be described as *direct* proofs in that they proceed from the given assumptions directly to the conclusion via a series of implications. We now discuss two other types of proof, both very commonly used.

The first is *proof by contradiction*. Suppose we wish to prove the truth of a statement P. A proof by contradiction would proceed by first assuming that P is false, i.e., assuming $\bar{P}$. We would try to deduce from this a statement Q that is palpably false (for example Q could be the statement "$0 = 1$" or "Liebeck is the Pope"). Having done this, we have shown

$$\bar{P} \Rightarrow Q \,.$$

Hence also $\bar{Q} \Rightarrow P$. Since we know Q is false, $\bar{Q}$ is true, and hence so is P, so we have proved P, as desired.

The next three examples illustrate the method of proof by contradiction.

Example 1.3
Let n be an integer such that n^2 is a multiple of 3. Then n is also a multiple of 3.

PROOF   Suppose n is not a multiple of 3. Then when we divide n by 3, we get a remainder of either 1 or 2; in other words, n is either 1 or 2 more than a multiple of 3. If the remainder is 1, then $n = 1 + 3k$ for some integer k, so

$$n^2 = (1 + 3k)^2 = 1 + 6k + 9k^2 = 1 + 3\left(2k + 3k^2\right) \,.$$

But this means that n^2 is 1 more than a multiple of 3, which is false, as we are given that n^2 is a multiple of 3. And if the remainder is 2, then $n = 2 + 3k$ for some integer k, so

$$n^2 = (2 + 3k)^2 = 4 + 12k + 9k^2 = 1 + 3\left(1 + 4k + 3k^2\right) \,,$$

which is again false as n^2 is a multiple of 3.

Thus we have shown that assuming n is not a multiple of 3 leads to a false statement. Hence, as explained above, we have proved that n is a multiple of 3. ∎

Note Usually in a proof by contradiction, when we arrive at our false statement Q, we simply write something like "this is a contradiction," and stop. We do this in the next proof.

Example 1.4
No real number has square equal to -1.

PROOF   Suppose the statement is false. This means that there *is* a real number, say x, such that $x^2 = -1$. However, it is a general fact about real numbers that the square of any real number is greater than or equal to 0 (see Chapter 4, Example 4.2). Hence $x^2 \geq 0$, which implies that $-1 \geq 0$. This is a contradiction.   ∎

Example 1.5
Prove that $\sqrt{2} + \sqrt{6} < \sqrt{15}$.

PROOF   Let me start by giving a non-proof:

$$\sqrt{2} + \sqrt{6} < \sqrt{15} \Rightarrow \left(\sqrt{2} + \sqrt{6}\right)^2 < 15$$
$$\Rightarrow 8 + 2\sqrt{12} < 15 \Rightarrow 2\sqrt{12} < 7 \Rightarrow 48 < 49 \, .$$

The last statement ($48 < 49$) is true, so why is this not a proof? Because the implication is going the wrong way — we have shown that if P is the statement we want to prove, and Q is the statement that $48 < 49$, then $P \Rightarrow Q$; but this tells us nothing about the truth or otherwise of P.

A cunning change to the above false proof gives a correct proof, by contradiction. So assume the result is false, i.e., that $\sqrt{2} + \sqrt{6} \geq \sqrt{15}$. Then

$$\sqrt{2} + \sqrt{6} \geq \sqrt{15} \Rightarrow \left(\sqrt{2} + \sqrt{6}\right)^2 \geq 15$$
$$\Rightarrow 8 + 2\sqrt{12} \geq 15 \Rightarrow 2\sqrt{12} \geq 7 \Rightarrow 48 \geq 49 \, ,$$

which is a contradiction. Hence we have proved that $\sqrt{2} + \sqrt{6} < \sqrt{15}$.   ∎

The other method of proof we shall discuss is actually a way of proving statements are false, i.e., *disproving* them. We call the method *disproof by counterexample*. It is best explained by examples:

Example 1.6

Consider the following two statements:

(a) All men are Chinese.

(b) Every positive integer is equal to the sum of two integer squares.

As the reader will have cleverly spotted, both these statements are false. To disprove (a), we need to prove the negation, which is "there exists a man who is not Chinese"; this is readily done by simply displaying one man who is not Chinese — this man will then be a *counterexample* to statement (a). The point is that to disprove (a), we do not need to consider *all* men, we just need to produce a single counterexample.

Likewise, to disprove (b) we just need to provide a single counterexample — that is, a positive integer which is *not* equal to the sum of two squares. The number 3 fits the bill nicely.

Exercises for Chapter 1

1. Let A be the set $\{\alpha, \{1, \alpha\}, \{3\}, \{\{1, 3\}\}, 3\}$. Which of the following statements are true and which are false?

   (a) $\alpha \in A$ (f) $\{\{1, 3\}\} \subseteq A$
   (b) $\{\alpha\} \notin A$ (g) $\{\{1, \alpha\}\} \subseteq A$
   (c) $\{1, \alpha\} \subseteq A$ (h) $\{1, \alpha\} \notin A$
   (d) $\{3, \{3\}\} \subseteq A$ (i) $\emptyset \subseteq A$
   (e) $\{1, 3\} \in A$

2. Which of the following arguments are valid? For the valid ones, write down the argument symbolically.

   (a) I eat chocolate if I am depressed. I am not depressed. Therefore I am not eating chocolate.

   (b) I eat chocolate only if I am depressed. I am not depressed. Therefore I am not eating chocolate.

   (c) If a movie is not worth seeing, then it was not made in England. A movie is worth seeing only if critic Ivor Smallbrain reviews it. The movie *The Good, the Bad and the Mathematician* was not reviewed by Ivor Smallbrain. Therefore *The Good, the Bad and the Mathematician* was not made in England.

3. Which of the following statements are true, and which are false?

   (a) $n = 3$ only if $n^2 - 2n - 3 = 0$.

   (b) $n^2 - 2n - 3 = 0$ only if $n = 3$.

   (c) If $n^2 - 2n - 3 = 0$ then $n = 3$.

   (d) For integers a and b, ab is a square only if both a and b are squares.

   (e) For integers a and b, ab is a square if both a and b are squares.

4. Write down careful proofs of the following statements:

   (a) $\sqrt{6} - \sqrt{2} > 1$.

   (b) If n is an integer such that n^2 is even, then n is even.

   (c) If $n = m^3 - m$ for some integer m, then n is a multiple of 6.

5. Disprove the following statements:

   (a) if n and k are positive integers, then $n^k - n$ is always divisible by k

    (b) every positive integer is the sum of three squares (the squares being 0, 1, 4, 9, 16, etc.)

6. Prove by contradiction that a real number which is less than every positive real number cannot be positive.

7. (For fun!) $a_1, a_2, a_3, \ldots, a_n, \ldots$ are positive integers such that for all $n \geq 1$,

$$a_{n+1} > a_n, \text{ and } a_{a_n} = 3n .$$

    (a) Find a_1. (*Hint:* Let $x = a_1$. Then what is a_x?)

    (b) Find $a_2, a_3, \ldots, a_9$.

    (c) Find a_{100}.

    (d) Investigate the sequence $a_1, a_2, \ldots, a_n, \ldots$ further.

Chapter 2

Number Systems

In this chapter we introduce three number systems: the real numbers, the integers and the rationals.

The Real Numbers

Here is an infinite straight line:

$\cdots$ ———————————————————— $\cdots$

Choose a point on this line and label it as 0. Also choose a unit of length, and use it to mark off evenly spaced points on the line, labelled by the whole numbers $\ldots, -2, -1, 0, 1, 2, \ldots$ like this:

$$\cdots \quad \begin{array}{ccccccc} -3 & -2 & -1 & 0 & 1 & 2 & 3 \end{array} \quad \cdots$$

We shall think of the real numbers as the points on this line. Viewed in this way, the line is called the *real line*. Write $\mathbb{R}$ for the set of all real numbers.

The real numbers have a natural *ordering,* which we now describe. If x and y are real numbers, we write $x < y$, or equivalently $y > x$, if x is to the left of y on the real line; under these circumstances we say x is less than y, or y is greater than x. Also, $x \leq y$ indicates that x is less than or equal to y. Thus, the following statements are all true: $1 \leq 1$, $1 \geq 1$, $1 < 2$, $2 \geq 1$. A real number x is *positive* if $x > 0$, and is *negative* if $x < 0$.

The *integers* are the whole numbers, marked as above on the real line. We write $\mathbb{Z}$ for the set of all integers, and $\mathbb{N}$ for the set of all positive integers $\{1, 2, 3, \ldots\}$. Positive integers are sometimes called *natural numbers.*

Fractions $\frac{m}{n}$ can also be marked on the real line. For example, $\frac{1}{2}$ is placed halfway between 0 and 1; in general, $\frac{m}{n}$ can be marked by dividing each of the unit intervals into n equal sections, and counting m of these sections away from 0. A real number of the form $\frac{m}{n}$ (where m, n are integers) is called a *rational number.* We write $\mathbb{Q}$ for the set of all rational numbers.

There are of course many different fractions representing the same rational number: for example, $\frac{8}{12} = \frac{-6}{-9} = \frac{2}{3}$, and so on. We say the rational $\frac{m}{n}$ is in *lowest terms* if no cancelling is possible, i.e., if m and n have no common factors (apart from 1 and -1).

Rationals can be added and multiplied according to the familiar rules:

$$\frac{m}{n} + \frac{p}{q} = \frac{mq + np}{nq}, \quad \frac{m}{n} \times \frac{p}{q} = \frac{mp}{nq}.$$

Notice that the sum and product of two rationals is again rational.

In fact, addition and multiplication of arbitrary real numbers can be defined in such a way as to obey the following rules:

RULES 2.1
For all $a, b, c, \in \mathbb{R}$,
  (1) $a + b = b + a$ *and* $ab = ba$
  (2) $a + (b + c) = (a + b) + c$ *and* $a(bc) = (ab)c$
  (3) $a(b + c) = ab + ac$.

For example, (2) assures us that $(2 + 5) + (-3) = 2 + (5 + (-3))$ (i.e., $7-3 = 2+2$), and $(2 \times 5) \times (-3) = 2 \times (5 \times (-3))$ (i.e., $10 \times (-3) = 2 \times (-15)$).

Before proceeding, let us pause briefly to reflect on these rules. They may seem "obvious" in some sense, in that you have probably been assuming them for years without thinking. But ponder the following equation, to be solved for x:

$$x + 3 = 5.$$

What are the steps we carry out when we solve this equation? Here they are:
  Step 1. Add -3 to both sides: $(x + 3) + (-3) = 5 + (-3)$.
  Step 2. Apply rule (2): $x + (3 + (-3)) = 5 - 3$.
  Step 3. This gives $x + 0 = 5 - 3$, hence $x = 2$.
The point is that without rule (2) we would be stuck. (Indeed, there are strange systems of objects with an addition for which one does not have rule (2), and in such systems one cannot even solve simple equations like the one above.)

There are some further important rules obeyed by the real numbers, relating to the ordering described above. We postpone discussion of these until Chapter 4.

Rationals and Irrationals

We often call a rational number simply a rational. The next result shows that the rationals are densely packed on the real line.

PROPOSITION 2.1
Between any two rationals there is another rational.

PROOF   Let r and s be two different rationals. Say r is the larger, so $r > s$. We claim that the real number $\frac{1}{2}(r + s)$ is a rational lying between r and s. To see this, observe that $\frac{1}{2}r > \frac{1}{2}s \Rightarrow \frac{1}{2}r + \frac{1}{2}s > \frac{1}{2}s + \frac{1}{2}s \Rightarrow \frac{1}{2}(r + s) > s$, and likewise $\frac{1}{2}r > \frac{1}{2}s \Rightarrow \frac{1}{2}r + \frac{1}{2}r > \frac{1}{2}s + \frac{1}{2}r \Rightarrow r > \frac{1}{2}(r + s)$. Thus, $\frac{1}{2}(r + s)$ lies between r and s. Finally, it is rational, since if $r = \frac{m}{n}, s = \frac{p}{q}$, then $\frac{1}{2}(r + s) = \frac{mq+np}{2nq}$.   ∎

Despite its innocent statement and quick proof, this is a rather significant result. For example, it implies that in contrast to the integers, there is no smallest positive rational, since for any positive rational x there is a smaller positive rational, for example $\frac{1}{2}x$; likewise, given any rational, there is no "next rational up." The proposition also shows that the rationals cannot be represented completely by "dots" on the real line, since between any two dots there would have to be another dot.

The proposition also raises a profound question: OK, the rationals are dense on the real line; but do they in fact fill out the whole line? In other words, is every real number a rational?

The answer is no, as we shall now demonstrate. First we need the following proposition, which is not quite as obvious as it looks.

PROPOSITION 2.2
There is a real number α such that $\alpha^2 = 2$.

PROOF   Draw a square of side 1:

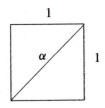

Let α be the length of a diagonal of the square. Then by Pythagoras, $\alpha^2 = 2$.

For the real number α in Proposition 2.2, we adopt the usual notation $\alpha = \sqrt{2}$.

PROPOSITION 2.3
$\sqrt{2}$ is not rational.

PROOF   This is a proof by contradiction. Suppose the statement is false, i.e., suppose $\sqrt{2}$ is rational. This means that there are integers m, n such that

$$\sqrt{2} = \frac{m}{n} .$$

Take $\frac{m}{n}$ to be in lowest terms (recall that this means that m, n have no common factors greater than 1).

Squaring the above equation gives $2 = \frac{m^2}{n^2}$, hence

$$m^2 = 2n^2 .$$

If m was odd, then m^2 would be odd by Example 1.2; but $m^2 = 2n^2$ is clearly even, so this cannot be the case. Therefore, m is even. Hence, we can write $m = 2k$, where k is an integer. Then

$$m^2 = 4k^2 = 2n^2 .$$

Consequently $n^2 = 2k^2$. So n^2 is even, and again by Example 1.2, this means n is also even.

We have now shown that both m and n are even. However, this means that the fraction $\frac{m}{n}$ is *not* in lowest terms. This is a contradiction. Therefore, $\sqrt{2}$ is not rational.   ∎

Note The following slightly more complicated geometrical argument than that given in Proposition 2.2 shows the existence of the real number $\sqrt{n}$ for any positive integer n. As in the figure below, draw a circle with diameter AB, with a point D marked so that $AD = n$, $DB = 1$. We leave it to the reader to use Pythagoras in the right-angled triangles ACD, BCD and ABC to show that the length CD has square equal to n, and hence $CD = \sqrt{n}$.

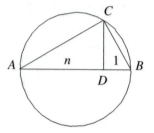

A real number which is not a rational is called an *irrational* number (or just an irrational). Thus $\sqrt{2}$ is an irrational, by Proposition 2.3. The next result enables us to construct many more examples of irrationals.

PROPOSITION 2.4

Let a be a rational number, and b an irrational.
  (i) Then $a + b$ is irrational.
  (ii) If $a \neq 0$, then ab is also irrational.

PROOF   (i) We prove this by contradiction. Suppose $a + b$ is rational, say $a + b = \frac{m}{n}$. Then writing the rational a as $\frac{p}{q}$, we have

$$b = \frac{m}{n} - a = \frac{m}{n} - \frac{p}{q} .$$

However, the right-hand side is rational, whereas b is given to be irrational, so this is a contradiction. Hence, $a + b$ is irrational.

The proof of part (ii) is very similar to that of (i), and we leave it to the reader. ∎

Example 2.1

The proposition shows that, for example, $1 + \sqrt{2}$ and $-5\sqrt{2}$ are irrational; indeed, $r + s\sqrt{2}$ is irrational for any rationals r, s with $s \neq 0$. Note also that there exist many further irrationals, not of this form. For instance, $\sqrt{3}$ is irrational, and there are no rationals r, s such that $\sqrt{3} = r + s\sqrt{2}$ (see Exercise 1 below).

Thus, there are "many" irrationals, in some sense. The next result shows that, like the rationals, the irrationals are densely packed on the real line.

PROPOSITION 2.5

Between any two real numbers there is an irrational.

PROOF   Let a and b be two real numbers, and say a is the smaller, so $a < b$. Choose a positive integer n which is larger than the real number $\frac{\sqrt{2}}{b-a}$. Then

$$\frac{\sqrt{2}}{n} < b - a .$$

If a is rational, then by Proposition 2.4, $a + \frac{\sqrt{2}}{n}$ is irrational; it also lies between a and b. And if a is irrational, then $a + \frac{1}{n}$ is irrational and lies between a and b. ∎

Exercises for Chapter 2

1. (a) Prove that $\sqrt{3}$ is irrational. (*Hint:* Example 1.3 should be useful.)

   (b) Prove that there are no rationals r, s such that $\sqrt{3} = r + s\sqrt{2}$.

2. (a) Prove that $\sqrt{2} + \sqrt{3}$ is irrational.

   (b) Prove that $\sqrt{2} + \sqrt{3} - \sqrt{5 + 2\sqrt{6}}$ is rational.

3. Show that the product of a non-zero rational and an irrational is always irrational.

4. (a) Let a, b be rationals and x irrational. Show that if $\frac{x+a}{x+b}$ is rational, then $a = b$.

   (b) Let x, y be rationals such that $\frac{x^2+x+\sqrt{2}}{y^2+y+\sqrt{2}}$ is also rational. Prove that either $x = y$ or $x + y = -1$.

5. Prove that if n is any positive integer, then $\sqrt{n} + \sqrt{2}$ is irrational.

6. Prove that between any two different real numbers there is a rational number and an irrational number.

7. In his review of the film *The Good, the Bad and the Mathematician*, critic Ivor Smallbrain wrote "I give this film a score of $\sqrt{n-2} + \sqrt{n+2}$ out of a million, where n is a positive integer." Given that Smallbrain's scores are always integers, how did he rate the film? (*Hint:* you may assume the following fact. The square root of a positive integer k is rational if and only if k is an integer square.)

Chapter 3

Decimals

It is all very well to have the real number system as points on the real line, but it is hard to prove any interesting facts about the reals without any convenient notation for them. We now remedy this by introducing the decimal notation for reals, and demonstrating a few of its basic properties.

We are all familiar with the following decimal expressions:

$$\frac{1}{2} = 0.50000\ldots$$

$$\frac{1}{9} = 0.11111\ldots$$

$$\frac{1}{7} = 0.142857142857\ldots$$

But what do we mean when we write, for example, $\frac{1}{9} = 0.11111\ldots$? We mean that the "sum to infinity" of the series

$$0.1111\ldots = 0.1 + 0.01 + 0.001 + \cdots = \frac{1}{10} + \frac{1}{10^2} + \frac{1}{10^3} + \cdots$$

is $\frac{1}{9}$; in other words, we can get as close as we like to $\frac{1}{9}$ provided we sum enough terms of the series. To make this absolutely precise would require us to go into the concepts of limits and convergence, which I do not intend to do in this book (it will presumably be part of your next course in Pure Mathematics!). Nevertheless, I hope the meaning is reasonably clear.

The above fact about $\frac{1}{9}$ is a special case of the following result on geometric series, which is probably very familiar.

PROPOSITION 3.1

Let x be a real number.
  (i) If $x \neq 1$, then $x + x^2 + x^3 + \cdots + x^n = \frac{x(1-x^n)}{1-x}$.
  (ii) If $-1 < x < 1$, then the sum to infinity

$$x + x^2 + x^3 + \cdots = \frac{x}{1-x}.$$

PROOF   (i) Let $s_n = x + x^2 + x^3 + \cdots + x^n$. Then $xs_n = x^2 + x^3 + \cdots + x^n + x^{n+1}$. Subtracting, we get $(1 - x)s_n = x - x^{n+1}$, which gives (i).

(ii) Since $-1 < x < 1$, we can make x^n as small as we like, provided we take n large enough. So we can make the sum in (i) as close as we like to $\frac{x}{1-x}$ provided we sum enough terms. That is to say, the sum to infinity is $\frac{x}{1-x}$. ∎

Putting $x = \frac{1}{10}$ in this proposition gives $\frac{1}{10} + \frac{1}{10^2} + \frac{1}{10^3} + \cdots = \frac{1}{9}$, as claimed above.

Likewise, in general the decimal expression $a_0.a_1a_2a_3\ldots$ where a_0 is an integer and $a_1, a_2, \ldots$ are integers between 0 and 9, means the real number which is the sum to infinity of the series

$$a_0 + \frac{a_1}{10} + \frac{a_2}{10^2} + \frac{a_3}{10^3} + \cdots$$

With this understanding, we obtain the next result, which gives us the convenient decimal notation for all real numbers.

PROPOSITION 3.2
Every real number x has a decimal expression

$$x = a_0.a_1a_2a_3\ldots .$$

PROOF   Picture x on the real line. Certainly x lies between two consecutive integers; let a_0 be the lower of these, so that

$$a_0 \leq x < a_0 + 1 .$$

Now divide the line between a_0 and $a_0 + 1$ into ten equal sections. Certainly x lies in one of these sections, so we can find a_1 between 0 and 9 such that

$$a_0 + \frac{a_1}{10} \leq x < a_0 + \frac{a_1 + 1}{10} .$$

Similarly we can find a_2 such that

$$a_0 + \frac{a_1}{10} + \frac{a_2}{10^2} \leq x < a_0 + \frac{a_1}{10} + \frac{a_2 + 1}{10^2} ,$$

and so on. If we do this enough times, the sum $a_0 + \frac{a_1}{10} + \frac{a_2}{10^2} + \cdots$ gets as close as we like to x. As explained above, this is what we mean by saying that $x = a_0.a_1a_2a_3\ldots .$ ∎

Example
We use the method of the proof just given to find the first few digits in the decimal expression for $\sqrt{2}$. Let $\sqrt{2} = a_0.a_1a_2a_3\ldots .$ First, observe that

$1^2 = 1$ and $2^2 = 4$, so $\sqrt{2}$ lies between 1 and 2, and hence $a_0 = 1$. Next, $(1.4)^2 = 1.96$ while $(1.5)^2 = 2.25$, so $a_1 = 4$. Likewise, $(1.41)^2 < 2$ while $(1.42)^2 > 2$, so $a_2 = 1$. We can continue finding decimal digits in this way until we get really fed up.

We now have a convenient notation for all real numbers: they all have decimal expressions. Two basic questions about this notation arise immediately:

(1) Can the same real number have two different decimal expressions; and if so, can we describe exactly when this happens?

(2) Which decimal expressions are rational and which are irrational?

We shall answer these questions in the next few results.

For (1), notice first that

$$0.9999\ldots = \frac{9}{10} + \frac{9}{10^2} + \frac{9}{10^3} + \cdots$$
$$= 9\left(\frac{1}{10} + \frac{1}{10^2} + \frac{1}{10^3} + \cdots\right) = 9\left(\frac{1}{9}\right) = 1.$$

Thus, the real number 1 has two different decimal expressions:

$$1 = 1.0000\ldots = 0.9999\ldots$$

Similarly, for example,

$$0.2579999\ldots = 0.2580000\ldots, \quad \text{and} \quad 1299.9999\ldots = 1300.0000\ldots,$$

and so on. Is this the only way two different decimal expressions can be equal? The answer is yes.

PROPOSITION 3.3

Suppose that $a_0.a_1a_2a_3\ldots$ and $b_0.b_1b_2b_3\ldots$ are two different decimal expressions for the same real number. Then one of these expressions ends in $9999\ldots$ and the other ends in $0000\ldots$.

PROOF Suppose first that $a_0 = b_0 = 0$. Call the real number with these two expressions x, so that

$$x = 0.a_1a_2a_3\ldots = 0.b_1b_2b_3\ldots \qquad (3.1)$$

Let the first place where the two expressions disagree be the kth place (k could be 1 of course). Thus $x = 0.a_1\ldots a_{k-1}a_k\ldots = 0.a_1\ldots a_{k-1}b_k\ldots$, where $a_k \neq b_k$. There is no harm in assuming $a_k > b_k$, hence $a_k \geq b_k + 1$. Then

$$x \geq 0.a_1\ldots a_{k-1}a_k000\ldots$$

and
$$x \leq 0.a_1 \ldots a_{k-1} b_k 999 \ldots = 0.a_1 \ldots a_{k-1}(b_k + 1)000 \ldots \, .$$

It follows that $a_k = b_k + 1$ and that the two expressions for x in (3.1) are $0.a_1 \ldots a_k 000 \ldots$ and $0.a_1 \ldots a_{k-1}(a_k - 1)999 \ldots .$

Finally, to handle the general case (where a_0, b_0 are not assumed to be 0), we replace a_0, b_0 with their expressions as integers using decimal digits, and apply the above argument. ∎

This provides us with a satisfactory answer to our question (1) above.

Now we address question (2): Which decimal expressions are rational and which are irrational? Choose a couple of rationals at random — say $\frac{8}{7}$ and $\frac{13}{22}$ — and work out their decimal expressions:

$$\frac{8}{7} = 1.142857142857\ldots, \quad \frac{13}{22} = 0.59090909 \ldots .$$

We observe that they have a striking feature in common: there is a sequence of digits which eventually repeats forever. We call such a decimal expression *periodic*.

In general, a periodic decimal is one that takes the form

$$a_0.a_1 \ldots a_k \, b_1 \ldots b_l \, b_1 \ldots b_l \, b_1 \ldots b_l \ldots .$$

We abbreviate this expression by writing it as $a_0.a_1 \ldots a_k \overline{b_1 \ldots b_l}$. The *period* of such a decimal is the number of digits in a repeating sequence of smallest length. For example, the decimal expression for $\frac{8}{7}$ has period 6.

The next result should not come as a major surprise.

PROPOSITION 3.4
The decimal expression for any rational number is periodic.

PROOF   Consider a rational $\frac{m}{n}$ (where $m, n \in \mathbb{Z}$). To express this as a decimal, we perform long division of n into $m.0000\ldots .$ At each stage of the long division, we get a remainder which is one of the n integers between 0 and $n - 1$. Therefore, eventually we must get a remainder which occurred before. The digits between the occurrences of these remainders will then repeat forever. ∎

Proposition 3.4 tell us that

$$a_0.a_1 a_2 a_3 \ldots \text{ rational} \Rightarrow a_0.a_1 a_2 a_3 \ldots \text{ periodic} \, .$$

It would be very nice if the reverse implication were also true, i.e., periodic $\Rightarrow$ rational. Let us first consider an example.

Example 3.1

Let $x = 0.3\overline{14}$. Is x rational? Well,

$$x = \frac{3}{10} + \frac{14}{10^3} + \frac{14}{10^5} + \frac{14}{10^7} + \cdots = \frac{3}{10} + \frac{14}{10^3}\left(1 + \frac{1}{10^2} + \frac{1}{10^4} + \cdots\right).$$

The series in the brackets is a geometric series, which by Proposition 3.1 has sum to infinity $\frac{100}{99}$, so

$$x = \frac{3}{10} + \frac{14}{10^3} \cdot \frac{100}{99} = \frac{311}{990}.$$

In particular, x is rational.

It is not at all hard to generalize this argument to show that the reverse implication (periodic $\Rightarrow$ rational) is indeed true:

PROPOSITION 3.5

Every periodic decimal is rational.

PROOF Let $x = a_0.a_1 \ldots a_k\overline{b_1 \ldots b_l}$ be a periodic decimal. Define

$$A = a_0.a_1 \ldots a_k, \quad B = 0.b_1 \ldots b_l .$$

Then A and B are both rationals, and

$$x = A + \frac{B}{10^k}\left(1 + \frac{1}{10^l} + \frac{1}{10^{2l}} + \cdots\right) = A + \frac{B}{10^k} \cdot \frac{10^l}{10^l - 1},$$

which is clearly also rational. ∎

Exercises for Chapter 3

1. Express the decimal $1.\overline{813}$ as a fraction $\frac{m}{n}$ (where m and n are integers).

2. Prove carefully that the real number $0.101001000100001000001\ldots$ is irrational.

3. Show that the decimal expression for $\sqrt{2}$ is not periodic.

4. Let $x = 0.a_1a_2a_3\ldots$, where for $n = 1, 2, 3, \ldots$, the value of a_n is the number 0, 1 or 2 which is the remainder on dividing n by 3. Is x rational? If so, express x as a fraction $\frac{m}{n}$.

5. Without using a calculator, find the cube root of 2, correct to 1 decimal place.

6. Critic Ivor Smallbrain is watching the classic film $11.\overline{9}$ *Angry Men*. But he is bored, and starts wondering idly exactly which rational numbers $\frac{m}{n}$ have decimal expressions ending in $0000\ldots$ (i.e., repeating zeroes). He notices that this is the case if the denominator n is 2, 4, 5, 8, 10 or 16, and wonders if there is a simple general rule which tells you which rationals have this property.

   Help Ivor by proving that a rational $\frac{m}{n}$ (in lowest terms) has decimal expression ending in repeating zeroes, if and only if the denominator n is of the form $2^a 5^b$, where $a, b \geq 0$ and a, b are integers.

Chapter 4

Inequalities

An *inequality* is a statement about real numbers involving one of the symbols
">," "$\geq$," "<" or "$\leq$," for example $x > 2$ or $x^2 - 4y \leq 2x + 2$. In this
chapter we shall present some elementary notions concerning manipulation of
inequalities.

Recall from Chapter 2 the basic Rules 2.1 satisfied by addition and multiplication of real numbers. As we mentioned there, there are various further rules
concerning the ordering of the real numbers. Here they are:

RULES 4.1

(1) *If $x \in \mathbb{R}$, then either $x > 0$ or $x < 0$ or $x = 0$ (and just one of these is
true).*

(2) *If $x > y$ then $-x < -y$.*

(3) *If $x > y$ and $c \in \mathbb{R}$, then $x + c > y + c$.*

(4) *If $x > 0$ and $y > 0$, then $xy > 0$.*

(5) *If $x > y$ and $y > z$ then $x > z$.*

The rest of the chapter consists of six examples showing how to use these
rules to manipulate inequalities.

Example 4.1
If $x < 0$ then $-x > 0$.

PROOF   Applying (2) with 0 instead of x, and x instead of y, we see that
$x < 0 \Rightarrow -x > 0$.   ∎

Example 4.2
If $x \neq 0$ then $x^2 > 0$.

PROOF   If $x > 0$ then by (4), $x^2 = xx > 0$. If $x < 0$ then $-x > 0$ (by Example 4.1), so (4) gives $(-x)(-x) > 0$, i.e., $x^2 > 0$.   ∎

Example 4.3
If $x > 0$ and $u > v$ then $xu > xv$.

PROOF   We have

$$u > v \Rightarrow u - v > v - v = 0 \quad \text{(by (3) with } c = -v)$$
$$\Rightarrow x(u - v) > 0 \quad \text{(by (4))}$$
$$\Rightarrow xu - xv > 0$$
$$\Rightarrow xu - xv + xv > xv \quad \text{(by (3) with } c = xv)$$
$$\Rightarrow xu > xv .$$ ∎

Example 4.4
If $x > 0$ then $\frac{1}{x} > 0$.

PROOF   If $\frac{1}{x} < 0$ then $\frac{-1}{x} > 0$ by Example 4.1, so by (4), $x.\frac{-1}{x} > 0$, i.e., $-1 > 0$, a contradiction. Therefore $\frac{1}{x} \geq 0$. Since $\frac{1}{x} \neq 0$, we conclude by Rule 4.1(1) that $\frac{1}{x} > 0$.   ∎

Example 4.5
Let $x_1, x_2, \ldots, x_n \in \mathbb{R}$, and suppose that k of these numbers are negative and the rest are positive. If k is even, then the product $x_1 x_2 \ldots x_n > 0$. And if k is odd, $x_1 x_2 \ldots x_n < 0$.

PROOF   Since the order of the x_is does not matter, we may as well assume that $x_1, \ldots, x_k$ are negative, and $x_{k+1}, \ldots, x_n$ are positive. Then by Example 4.1, $-x_1, \ldots, -x_k, x_{k+1}, \ldots, x_n$ are all positive. By (4), the product of all of these is positive, so

$$(-1)^k x_1 x_2, \ldots, x_n > 0 .$$

If k is even this says $x_1 x_2, \ldots, x_n > 0$. And if k is odd it says $-x_1 x_2, \ldots, x_n > 0$, hence $x_1 x_2, \ldots, x_n < 0$.   ∎

The next example is a typical elementary inequality to solve:

Example 4.6
For which values of x is $x < \frac{2}{x+1}$?

Answer First, a word of warning — we cannot multiply both sides by $x + 1$, as this may or may not be positive. So we proceed more cautiously. Subtracting $\frac{2}{x+1}$ from both sides gives the inequality $x - \frac{2}{x+1} < 0$, which is the same as $\frac{x^2+x-2}{x+1} < 0$, i.e.,

$$\frac{(x+2)(x-1)}{x+1} < 0.$$

By Example 4.5, this is true if and only if either one or three of the quantities $x + 2$, $x - 1$, $x + 1$ is negative. All three are negative when $x < -2$, and just one is negative when $-1 < x < 1$.

Example 4.7
Show that $x^2 + x + 1 > 0$ for all $x \in \mathbb{R}$.

Answer Note that $x^2 + x + 1 = (x + \frac{1}{2})^2 + \frac{3}{4}$, hence using Example 4.2 and Rule 4.1(3), we have $x^2 + x + 1 \geq \frac{3}{4}$ for all x.

For a real number x, we define the *modulus* of x, written $|x|$, by

$$|x| = \begin{array}{l} x, \text{ if } x \geq 0 \\ -x, \text{ if } x < 0 \end{array}$$

For example, $|-5| = 5$, and $|7| = 7$. Notice that $|x|$ just measures the distance from the point x on the real line to the origin 0. Thus, for example, the set of values of x such that $|x| \leq 2$ consists of all x between -2 and 2, which we summarize as $-2 \leq x \leq 2$. More generally:

Example 4.8
Let $a, b \in \mathbb{R}$ with $b > 0$. For which values of x is the inequality $|x - a| \leq b$ satisfied?

Answer When $x \geq a$, the inequality says $x - a \leq b$, i.e., $x \leq a + b$. And when $x < a$, the inequality says $a - x \leq b$, i.e., $x \geq a - b$. So the range of values of x satisfying the inequality is $a - b \leq x \leq a + b$.

Example 4.9
Find all values of x such that $|x - 3| < 2|x + 3|$.

Answer We must be quite careful with this — the inequality varies according to whether $x < -3$, $-3 \leq x < 3$, or $x \geq 3$.

When $x < -3$ the inequality says $-(x - 3) < 2(-x - 3)$, which is the same as $x < -9$. When $-3 \leq x < 3$, the inequality says $-(x - 3) < 2(x + 3)$, i.e., $3x > -3$, in other words $x > -1$. And when $x > 3$ the inequality says

$x - 3 < 2(x + 3)$, i.e., $x > -9$. We deduce that the values of x satisfying the inequality are

$$x < -9 \quad \text{and} \quad x > -1.$$

Exercises for Chapter 4

1. Using Rules 4.1, show that if $x > 0$ and $y < 0$ then $xy < 0$, and that if $a > b > 0$ then $\frac{1}{a} < \frac{1}{b}$.

2. For which values of x is $x^2 + x + 1 \geq \frac{x-1}{2x-1}$?

3. For which values of x is $-3x^2 + 4x > 1$?

4. Prove that if $0 < u < 1$ and $0 < v < 1$ then $\frac{u+v}{1+uv} < 1$. For which other values of u, v is this inequality true?

5. Find the range of values of x such that

   (i) $|x + 5| \geq 1$

   (ii) $|x + 5| > |x - 2|$

   (iii) $|x + 5| < |x^2 + 2x + 3|$.

Chapter 5

n^{th} Roots and Rational Powers

In Chapter 2, just after proving Proposition 2.2, we gave a cunning geometrical construction which demonstrated the existence of the real number $\sqrt{n}$ for any positive integer n. However, proving the existence of a cube root, and more generally, an nth root, of any positive real number x is much harder, and requires a deeper analysis of the reals than we have undertaken thus far. We shall carry out such an analysis later, in Chapter 10. However, because we wish to include n^{th} roots in the discussion of complex numbers in the next chapter, we pick out the main result from Chapter 10 on such matters, namely Proposition 10.1, and state it here. (It is, of course, proved in Chapter 10.)

PROPOSITION 5.1
Let n be a positive integer. If x is a positive real number, then there is exactly one positive real number y such that $y^n = x$.

If x, y are as in the statement, we adopt the familiar notation

$$y = x^{\frac{1}{n}} .$$

Thus, for example, $5^{\frac{1}{2}}$ is the positive square root of 5, and $5^{\frac{1}{7}}$ is the unique positive real number y such that $y^7 = 5$.

We can extend this notation to define rational powers of positive reals, as follows. Let $x > 0$. Integer powers x^m ($m \in \mathbb{Z}$) are defined in the familiar way: if $m > 0$ then $x^m = xx \ldots x$, the product of m copies of x; if $m < 0$ then $x^m = \frac{1}{x^{-m}}$; and for $m = 0$ we define $x^0 = 1$.

Now let $\frac{m}{n} \in \mathbb{Q}$ (with $m, n \in \mathbb{Z}$ and $n \geq 1$), Then we define

$$x^{\frac{m}{n}} = \left(x^{\frac{1}{n}} \right)^m .$$

For example, $5^{-\frac{4}{7}}$ is defined to be $(5^{\frac{1}{7}})^{-4}$.

The basic rules concerning products of these rational powers are given in the next proposition. Although they probably seem rather familiar, they are not totally obvious, and require careful proof.

PROPOSITION 5.2
Let x, y be positive real numbers, and $p, q \in \mathbb{Q}$. Then

(i) $x^p x^q = x^{p+q}$

(ii) $(x^p)^q = x^{pq}$

(iii) $(xy)^p = x^p y^p$.

PROOF (i) We first establish the result when p and q are both integers. In this case, when $p, q \geq 0$, we have $x^p = x \ldots x$ (p factors), $x^q = x \ldots x$ (q factors), so

$$x^p x^q = (x \ldots x).(x \ldots x) = x^{p+q} ,$$

and when $p \geq 0, q < 0$, $x^q = 1/x \ldots x$ ($-q$ factors), so

$$x^p x^q = (x \ldots x)/(x \ldots x) = x^{p-(-q)} = x^{p+q} .$$

Similar arguments cover the other possibilities $p < 0, q \geq 0$ and $p, q < 0$.

Now let us consider the general case, where p, q are rationals. Write $p = \frac{m}{n}, q = \frac{h}{k}$ with $m, n, h, k \in \mathbb{Z}$. Then

$$x^p x^q = x^{\frac{m}{n}} x^{\frac{h}{k}} = x^{\frac{mk}{nk}} x^{\frac{hn}{nk}} = \left(x^{\frac{1}{nk}}\right)^{mk} \left(x^{\frac{1}{nk}}\right)^{hn} .$$

By the integer case of part (i), established in the previous paragraph, this is equal to

$$\left(x^{\frac{1}{nk}}\right)^{mk+hn} ,$$

which, by our definition of rational powers, is equal to

$$x^{\frac{mk+hn}{nk}} = x^{\frac{m}{n}+\frac{h}{k}} = x^{p+q} .$$

(ii, iii) First, as in (i), we easily establish the results for $p, q \in \mathbb{Z}$. Now let $p = \frac{m}{n}, q = \frac{h}{k}$ (with $h, k, m, n \in \mathbb{Z}$). By Proposition 5.1, there is a real number a such that $x = a^{nk}$. Then

$$(x^p)^q = \left(\left(a^{nk}\right)^{\frac{m}{n}}\right)^{\frac{h}{k}} = \left(\left(\left(\left(a^k\right)^n\right)^{\frac{1}{n}}\right)^m\right)^{\frac{h}{k}} = \left(\left(a^k\right)^m\right)^{\frac{h}{k}} = \left(\left(a^m\right)^k\right)^{\frac{1}{k}}^h$$

$$= \left(a^m\right)^h = a^{mh} = \left(\left(x^{1/nk}\right)^{mh}\right) = x^{\frac{mh}{nk}} = x^{pq} ,$$

and

$$x^p y^p = \left(x^{\frac{1}{n}}\right)^m \left(y^{\frac{1}{n}}\right)^m = \left(x^{\frac{1}{n}} y^{\frac{1}{n}}\right)^m = \left(\left(\left(x^{\frac{1}{n}} y^{\frac{1}{n}}\right)^n\right)^{\frac{1}{n}}\right)^m$$

$$= \left(\left(x^{\frac{1}{n}}\right)^n \left(y^{\frac{1}{n}}\right)^n\right)^{\frac{m}{n}} = (xy)^p . \quad \blacksquare$$

Exercises for Chapter 5

1. Show that $(50)^{3/4}(\frac{5}{\sqrt{2}})^{-1/2} = 10$.

2. What is the real cube root of $3^{(3^{333})}$?

3. Find all real solutions x of the equation $x^{1/2} - (2 - 2x)^{1/2} = 1$.

4. Let x and y be integers greater than 1, satisfying the equation $y^{4/3} = x^{7/6}$. Find the smallest such integers x, y.

Chapter 6

Complex Numbers

We all know that there are simple quadratic equations, such as $x^2 + 1 = 0$, which have no real solutions. In order to provide a notation with which to discuss such equations, we introduce a symbol i, and define

$$i^2 = -1 .$$

A *complex number* is defined to be a symbol $a + bi$, where a, b are real numbers. If $z = a + bi$, we call a the *real part* of z and b the *imaginary part,* and write

$$a = Re(z), \quad b = Im(z) .$$

We define addition and multiplication of complex numbers by the rules

addition: $(a + bi) + (c + di) = a + c + (b + d)i$

multiplication: $(a + bi)(c + di) = ac - bd + (ad + bc)i .$

Notice that in the multiplication rule, we multiply out the brackets in the usual way, and replace the i^2 by -1. For example, $(1 + 2i)(3 - i) = 5 + 5i$; and $(a + bi)(a - bi) = a^2 + b^2$.

It is also possible to subtract complex numbers:

$$(a + bi) - (c + di) = a - c + (b - d)i$$

and, less obviously, to divide them: provided c, d are not both 0,

$$\frac{a + bi}{c + di} = \frac{(a + bi)(c - di)}{(c + di)(c - di)} = \frac{ac + bd}{c^2 + d^2} + \left(\frac{bc - ad}{c^2 + d^2}\right)i .$$

For example, $\frac{1-i}{1+i} = \frac{(1-i)(1-i)}{(1+i)(1-i)} = \frac{-2i}{2} = -i$.

We write $\mathbb{C}$ for the set of all complex numbers. If we identify the complex number $a + 0i$ with the real number a, we see that $\mathbb{R} \subseteq \mathbb{C}$.

Notice that every quadratic equation $ax^2 + bx + c = 0$ (where $a, b, c \in R$) has roots in $\mathbb{C}$. For by the famous formula you will be familiar with, the roots are

$$\frac{1}{2a}\left(-b \pm \sqrt{b^2 - 4ac}\right) .$$

If $b^2 \geq 4ac$ these roots lie in $\mathbb{R}$, while if $b^2 < 4ac$ they are the complex numbers $\frac{-b}{2a} \pm \frac{\sqrt{4ac - b^2}}{2a} i$.

Geometrical Representation of Complex Numbers

It turns out to be a very fruitful idea to represent complex numbers by points in the xy-plane. This is done in a natural way — the complex number $a + bi$ is represented by the point in the plane with coordinates (a, b). For example, i is represented by $(0, 1)$; $1 - i$ by $(1, -1)$, and so on:

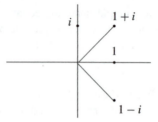

If $z = a + bi$, we define $\bar{z} = a - bi$, and call this the *complex conjugate* of z. Also, the *modulus* of $z = a + bi$ is the distance from the origin to the point (a, b) representing z. It is written as $|z|$. Thus,

$$|z| = \sqrt{a^2 + b^2} .$$

Notice that

$$z\bar{z} = (a + bi)(a - bi) = a^2 + b^2 = |z|^2 .$$

The *argument* of z is the angle θ between the x-axis and the line joining 0 to z, measured in the counterclockwise direction:

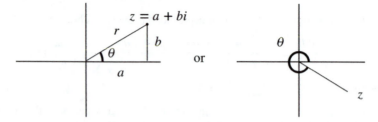

If $z = a + bi$ and $|z| = r$, then we see that $a = r \cos \theta$, $b = r \sin \theta$, so

$$z = r(\cos \theta + i \sin \theta) .$$

This is known as the *polar form* of the complex number z.

Example

The polar forms of $i, -1, 1 + i$ and $1 - i$ are

$$i = 1 \left(\cos \tfrac{\pi}{2} + i \sin \tfrac{\pi}{2}\right), \quad -1 = 1(\cos \pi + i \sin \pi),$$
$$1 + i = \sqrt{2} \left(\cos \tfrac{\pi}{4} + i \sin \tfrac{\pi}{4}\right), \quad 1 - i = \sqrt{2} \left(\cos \tfrac{7\pi}{4} + i \sin \tfrac{7\pi}{4}\right).$$

Let $z = r(\cos \theta + i \sin \theta)$. Notice that

$$\cos \theta + i \sin \theta = \cos(\theta + 2\pi) + i \sin(\theta + 2\pi)$$
$$= \cos(\theta + 4\pi) + i \sin(\theta + 4\pi) = \dots,$$

so multiples of 2π can be added to θ (or subtracted from θ) without changing z. Thus, z has many different arguments. There is, however, a unique value of the argument of z in the range $-\pi < \theta \le \pi$, and this is called the *principal argument* of z, written $arg(z)$. For example, $arg(1 - i) = -\tfrac{\pi}{4}$.

De Moivre's Theorem

The xy-plane, representing the set of complex numbers as just described, is known as the *Argand diagram;* it is also sometimes called simply the *complex plane.*

The significance of the geometrical representation of complex numbers begins to become apparent in the next result, which shows that complex multiplication has a simple and natural geometric interpretation.

THEOREM 6.1 De Moivre's Theorem

Let z_1, z_2 be complex numbers with polar forms

$$z_1 = r_1 (\cos \theta_1 + i \sin \theta_1), \quad z_2 = r_2 (\cos \theta_2 + i \sin \theta_2).$$

Then the product

$$z_1 z_2 = r_1 r_2 (\cos (\theta_1 + \theta_2) + i \sin (\theta_1 + \theta_2)).$$

In other words, $z_1 z_2$ has modulus $r_1 r_2$ and argument $\theta_1 + \theta_2$.

PROOF We have

$$z_1 z_2 = r_1 r_2 (\cos \theta_1 + i \sin \theta_1) (\cos \theta_2 + i \sin \theta_2)$$
$$= r_1 r_2 (\cos \theta_1 \cos \theta_2 - \sin \theta_1 \sin \theta_2 + i (\cos \theta_1 \sin \theta_2 + \sin \theta_1 \cos \theta_2))$$
$$= r_1 r_2 (\cos (\theta_1 + \theta_2) + i \sin (\theta_1 + \theta_2)). \quad \blacksquare$$

De Moivre's Theorem says that multiplying a complex number z by $\cos\theta + i\sin\theta$ rotates z counterclockwise through the angle θ; for example, multiplication by i rotates z through $\frac{\pi}{2}$:

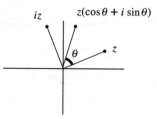

We now deduce a significant consequence of De Moivre's Theorem.

PROPOSITION 6.1
Let $z = r(\cos\theta + i\sin\theta)$, and let n be a positive integer. Then
(i) $z^n = r^n(\cos n\theta + i\sin n\theta)$, and
(ii) $z^{-n} = r^{-n}(\cos n\theta - i\sin n\theta)$.

PROOF   (i) Applying Theorem 6.1 with $z_1 = z_2 = z$ gives

$$z^2 = zz = rr(\cos(\theta + \theta) + i\sin(\theta + \theta)) = r^2(\cos 2\theta + i\sin 2\theta).$$

Repeating, we get

$$z^n = r \ldots r(\cos(\theta + \cdots + \theta) + i\sin(\theta + \cdots + \theta)) = r^n(\cos n\theta + i\sin n\theta).$$

(ii) First observe that

$$z^{-1} = \frac{1}{z} = \frac{1}{r(\cos\theta + i\sin\theta)} = \frac{1}{r}\frac{\cos\theta - i\sin\theta}{(\cos\theta + i\sin\theta)(\cos\theta - i\sin\theta)}$$
$$= \frac{1}{r}(\cos\theta - i\sin\theta).$$

Hence $z^{-1} = r^{-1}(\cos(-\theta) + i\sin(-\theta))$, which proves the result for $n = 1$. And, for general n, we simply note that $z^{-n} = (z^{-1})^n$, which by part (i) is equal to $(r^{-1})^n(\cos(-n\theta) + i\sin(-n\theta))$, hence to $(r^{-n})(\cos(n\theta) - i\sin(n\theta))$.   ∎

We now give a few examples illustrating the power of De Moivre's Theorem.

Example 6.1
Calculate $(-\sqrt{3} + i)^7$.

Answer We first find the polar form of $z = -\sqrt{3} + i$.

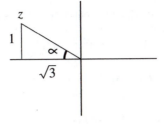

In the diagram, $\sin \alpha = \frac{1}{2}$, so $\alpha = \frac{\pi}{6}$. Hence $arg(z) = \frac{5\pi}{6}$. Also $|z| = 2$, so the polar form of z is

$$z = 2 \left(\cos \frac{5\pi}{6} + i \sin \frac{5\pi}{6} \right) .$$

Hence, by Proposition 6.1,

$$\left(-\sqrt{3} + i \right)^7 = 2^7 \left(\cos \frac{35\pi}{6} + i \sin \frac{35\pi}{6} \right)$$

$$= 2^7 \left(\cos \frac{-\pi}{6} + i \sin \frac{-\pi}{6} \right)$$

(subtracting 6π from the argument $\frac{35\pi}{6}$). Since $\cos \frac{-\pi}{6} = \frac{\sqrt{3}}{2}$ and $\sin \frac{-\pi}{6} = -\frac{1}{2}$, this gives

$$\left(-\sqrt{3} + i \right)^7 = 2^6 \left(\sqrt{3} - i \right) .$$

Example 6.2
Find a complex number w such that $w^2 = -\sqrt{3} + i$ (i.e., find a complex square root of $-\sqrt{3} + i$).

Answer From the previous solution, $-\sqrt{3} + i = 2(\cos \frac{5\pi}{6} + i \sin \frac{5\pi}{6})$. So if we define

$$w = \sqrt{2} \left(\cos \frac{5\pi}{12} + i \sin \frac{5\pi}{12} \right) ,$$

then by Proposition 6.1, $w^2 = -\sqrt{3} + i$. Note that $\sqrt{2}(\cos(\frac{5\pi}{12} + \pi) + i \sin(\frac{5\pi}{12} + \pi))$ works equally well; by Theorem 6.1 this is equal to $w(\cos \pi + i \sin \pi) = -w$.

Example 6.3
In this example we find a formula for $\cos 3\theta$ in terms of $\cos \theta$.
    We begin with the equation

$$\cos 3\theta + i \sin 3\theta = (\cos \theta + i \sin \theta)^3 .$$

Writing $c = \cos\theta$, $s = \sin\theta$, and expanding the cube, we get

$$\cos 3\theta + i\sin 3\theta = c^3 + 3c^2 si + 3cs^2 i^2 + s^3 i^3 = c^3 - 3cs^2 + i\left(3c^2 s - s^3\right).$$

Equating real parts, we have $\cos 3\theta = c^3 - 3cs^2$. Also $c^2 + s^2 = \cos^2\theta + \sin^2\theta = 1$, so $s^2 = 1 - c^2$, and therefore

$$\cos 3\theta = c^3 - 3c\left(1 - c^2\right) = 4c^3 - 3c.$$

That is,

$$\cos 3\theta = 4\cos^3\theta - 3\cos\theta.$$

Example 6.4

We now use the previous example to find a cubic equation having $\cos\frac{\pi}{9}$ as a root.

Putting $\theta = \frac{\pi}{9}$ and $c = \cos\frac{\pi}{9}$, Example 6.3 gives

$$\cos 3\theta = 4c^3 - 3c.$$

However, $\cos 3\theta = \cos\frac{\pi}{3} = \frac{1}{2}$. Hence $\frac{1}{2} = 4c^3 - 3c$. In other words, $c = \cos\frac{\pi}{9}$ is a root of the cubic equation

$$8x^3 - 6x - 1 = 0.$$

Note that if $\phi = \frac{\pi}{9} + \frac{2\pi}{3}$ or $\frac{\pi}{9} + \frac{4\pi}{3}$, then $\cos 3\phi = \frac{1}{2}$, and hence the above argument shows $\cos\phi$ is also a root of this cubic equation. The roots of $8x^3 - 6x - 1 = 0$ are therefore $\cos\frac{\pi}{9}$, $\cos\frac{7\pi}{9}$ and $\cos\frac{13\pi}{9}$.

The $e^{i\theta}$ Notation

It is somewhat cumbersome to keep writing $\cos\theta + i\sin\theta$ in our notation for complex numbers. We therefore introduce a rather more compact notation by defining

$$e^{i\theta} = \cos\theta + i\sin\theta$$

for any real number θ. (This equation turns out to be very significant when $e^{i\theta}$ is regarded as an exponential function, but for now it is simply the definition of the symbol $e^{i\theta}$.)

For example,

$$e^{2\pi i} = 1, \quad e^{\pi i} = -1, \quad e^{\frac{\pi}{2}i} = i, \quad e^{\frac{\pi}{4}i} = \frac{1}{\sqrt{2}}(1 + i).$$

Also, for any integer k,

$$e^{i\theta} = e^{i(\theta + 2k\pi)} .$$

Each of the complex numbers $e^{i\theta}$ has modulus 1, and the set consisting of all of them is the *unit circle* in the Argand diagram, i.e., the circle of radius 1 centered at the origin:

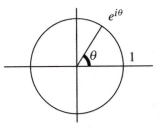

The polar form of a complex number z can now be written as

$$z = re^{i\theta}$$

where $r = |z|$ and $\theta = arg(z)$. For example,

$$-\sqrt{3} + i = 2e^{\frac{5\pi i}{6}} .$$

De Moivre's Theorem 6.1 implies that

$$e^{i\theta} e^{i\phi} = e^{i(\theta + \phi)} ,$$

and Proposition 6.1 says that for any integer n,

$$\left(e^{i\theta} \right)^n = e^{in\theta} .$$

From these facts we begin to see some of the significance behind the definition of $e^{i\theta}$ emerging.

PROPOSITION 6.2
(i) If $z = re^{i\theta}$ then $\bar{z} = re^{-i\theta}$.
(ii) Let $z = re^{i\theta}$, $w = se^{i\phi}$ in polar form. Then

$$z = w \iff r = s \text{ and } \theta - \phi = 2k\pi \text{ with } k \in \mathbb{Z} .$$

PROOF   (i) We have $z = r(\cos\theta + i\sin\theta)$, so $\bar{z} = r(\cos\theta - i\sin\theta) = r(\cos(-\theta) + i\sin(-\theta)) = re^{-i\theta}$.
   (ii) If $r = s$ and $\theta - \phi = 2k\pi$ with $k \in \mathbb{Z}$, then

$$z = re^{i\theta} = se^{i(\phi + 2k\pi)} = se^{i\phi} = w .$$

This does the "right to left" implication.

For the "left to right" implication, suppose $z = w$. Then $|z| = |w|$, so $r = s$ and also $e^{i\theta} = e^{i\phi}$. Now

$$e^{i\theta} = e^{i\phi} \Rightarrow e^{i\theta}e^{-i\phi} = e^{i\phi}e^{-i\phi} \Rightarrow e^{i(\theta - \phi)} = 1$$
$$\Rightarrow \cos(\theta - \phi) = 1, \quad \sin(\theta - \phi) = 0 \Rightarrow \theta - \phi = 2k\pi \text{ with } k \in \mathbb{Z} . \quad \blacksquare$$

Roots of Unity

Consider the equation

$$z^3 = 1 .$$

This is easy enough to solve: rewriting it as $z^3 - 1 = 0$, and factorizing this as $(z - 1)(z^2 + z + 1) = 0$, we see that the roots are

$$1, \quad -\frac{1}{2} + \frac{\sqrt{3}}{2}i, \quad -\frac{1}{2} - \frac{\sqrt{3}}{2}i .$$

These complex numbers have polar forms

$$1, \quad e^{\frac{2\pi i}{3}}, \quad e^{\frac{4\pi i}{3}} .$$

In other words, they are evenly spaced on the unit circle like this:

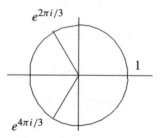

These three complex numbers are called the *cube roots of unity*.

More generally, if n is a positive integer, then the complex numbers which satisfy the equation

$$z^n = 1$$

are called the nth *roots of unity*.

PROPOSITION 6.3

Let n be a positive integer, and define $w = e^{\frac{2\pi i}{n}}$. Then the nth roots of unity are the n complex numbers

$$1, w, w^2, \ldots, w^{n-1}$$

(i.e., $1, e^{\frac{2\pi i}{n}}, e^{\frac{4\pi i}{n}}, \ldots, e^{\frac{2(n-1)\pi i}{n}}$). *They are evenly spaced around the unit circle.*

PROOF   Let $z = re^{i\theta}$ be an nth root of unity. Then

$$1 = z^n = r^n e^{ni\theta} .$$

From Proposition 6.2(ii) it follows that $r = 1$ and $n\theta = 2k\pi$ with $k \in \mathbb{Z}$. Therefore, $\theta = \frac{2k\pi}{n}$, and so $z = e^{\frac{2k\pi i}{n}} = w^k$.

Thus every n^{th} root of unity is a power of w. On the other hand, any power w^k is an nth root of unity, since

$$\left(w^k\right)^n = w^{nk} = \left(e^{\frac{2\pi i}{n}}\right)^{nk} = \left(e^{2\pi i}\right)^k = 1 .$$

The complex numbers

$$1, w, w^2, \ldots, w^{n-1}$$

are all the distinct powers of w (since $w^n = 1$, $w^{n+1} = w$, etc.). Hence, these are the nth roots of unity.   ∎

Example 6.5

The fourth roots of unity are $1, e^{\frac{i\pi}{2}}, e^{i\pi}, e^{\frac{3i\pi}{2}}$, which are just

$$1, i, -1, -i .$$

The 6th roots of unity are $1, e^{\frac{i\pi}{3}}, e^{\frac{2i\pi}{3}}, -1, e^{\frac{4i\pi}{3}}$ and $e^{\frac{5i\pi}{3}}$; these are the corners of a regular hexagon drawn inside the unit circle.

We can use the nth roots of unity to find the nth roots of any complex number. Here is an example.

Example 6.6

Find all solutions of the equation

$$z^5 = -\sqrt{3} + i .$$

(In other words, find all the fifth roots of $-\sqrt{3} + i$.)

Answer Let $p = -\sqrt{3} + i$. Recall that $p = 2e^{\frac{5\pi i}{6}}$. One of the fifth roots of this is clearly

$$\alpha = 2^{\frac{1}{5}} e^{\frac{\pi i}{6}}$$

(where of course $2^{\frac{1}{5}}$ is the real fifth root of 2). If w is a fifth root of unity, then $(\alpha w)^5 = \alpha^5 w^5 = \alpha^5 = z$, so αw is also a fifth root of p. Thus we have found the following 5 fifth roots of $-\sqrt{3} + i$:

$$\alpha, \alpha e^{\frac{2\pi i}{5}}, \alpha e^{\frac{4\pi i}{5}}, \alpha e^{\frac{6\pi i}{5}}, \alpha e^{\frac{8\pi i}{5}} .$$

These are in fact all the fifth roots of p: for if β is any fifth root of p, then $\beta^5 = \alpha^5 = p$, so $(\frac{\beta}{\alpha})^5 = 1$, which means that $\frac{\beta}{\alpha} = w$ is a fifth root of unity, and hence $\beta = \alpha w$ is in the above list.

We conclude that the fifth roots of $-\sqrt{3} + i$ are

$$2^{\frac{1}{5}} e^{\frac{\pi i}{6}}, \; 2^{\frac{1}{5}} e^{\frac{17\pi i}{30}}, \; 2^{\frac{1}{5}} e^{\frac{29\pi i}{30}}, \; 2^{\frac{1}{5}} e^{\frac{41\pi i}{30}}, \; 2^{\frac{1}{5}} e^{\frac{53\pi i}{30}} \; .$$

In general, the above method shows that if one of the nth roots of a complex number is β, then the others are $\beta w, \beta w^2, \ldots, \beta w^{n-1}$ where $w = e^{\frac{2\pi i}{n}}$.

Exercises for Chapter 6

1. Prove the following facts about complex numbers:

   (a) $u + v = v + u$ and $uv = vu$ for all $u, v \in \mathbb{C}$.

   (b) $\overline{u + v} = \bar{u} + \bar{v}$ and $\overline{uv} = \bar{u}\bar{v}$ for all $u, v \in \mathbb{C}$.

   (c) $|u|^2 = u\bar{u}$ for all $u \in \mathbb{C}$.

   (d) $u(vw) = (uv)w$ for all $u, v, w \in \mathbb{C}$.

   (*Hint:* for (d), use the polar forms of u, v, w.)

2. Find the real and imaginary parts of $(\sqrt{3} - i)^{10}$ and $(\sqrt{3} - i)^{-7}$. For which values of n is $(\sqrt{3} - i)^n$ real?

3. Find the seven roots of the equation $z^7 - \sqrt{3} + i = 0$. Which one of these roots is closest to the imaginary axis?

4. Express $\frac{1+i}{\sqrt{3}+i}$ in the form $x + iy$, where $x, y \in \mathbb{R}$. By writing each of $1 + i$ and $\sqrt{3} + i$ in polar form, deduce that

$$\cos \frac{\pi}{12} = \frac{\sqrt{3}+1}{2\sqrt{2}}, \quad \sin \frac{\pi}{12} = \frac{\sqrt{3}-1}{2\sqrt{2}}.$$

5. (a) Show that $x^5 - 1 = (x - 1)(x^4 + x^3 + x^2 + x + 1)$. Deduce that if $\omega = e^{2\pi i/5}$ then $\omega^4 + \omega^3 + \omega^2 + \omega + 1 = 0$.

   (b) Let $\alpha = 2\cos\frac{2\pi}{5}$ and $\beta = 2\cos\frac{4\pi}{5}$. Show that $\alpha = \omega + \omega^4$ and $\beta = \omega^2 + \omega^3$. Find a quadratic equation with roots α, β. Hence show that

$$\cos \frac{2\pi}{5} = \frac{1}{4}\left(\sqrt{5} - 1\right).$$

6. Find a formula for $\cos 4\theta$ in terms of $\cos \theta$. Hence write down a quartic equation (i.e., an equation of degree 4) which has $\cos \frac{\pi}{12}$ as a root. What are the other roots of your equation?

7. Find all complex numbers z such that $|z| = |\sqrt{2} + z| = 1$. Prove that each of these satisfies $z^8 = 1$.

8. Show that if w is an nth root of unity, then $\bar{w} = \frac{1}{w}$. Deduce that

$$\overline{(1 - w)^n} = (w - 1)^n.$$

Hence show that $(1 - w)^{2n}$ is real.

9. Let n be a positive integer, and let $z \in \mathbb{C}$ satisfy the equation

$$(z - 1)^n + (z + 1)^n = 0 .$$

   (a) Show that $z = \frac{1+w}{1-w}$ for some $w \in \mathbb{C}$ such that $w^n = -1$.

   (b) Show that $w\bar{w} = 1$.

   (c) Deduce that z lies on the imaginary axis.

10. Critic Ivor Smallbrain is discussing the film *Sets, Lines and Videotape* with his two chief editors, Sir Giles Tantrum and Lord Overthetop. They are sitting at a circular table of radius 1. Ivor is bored, and notices in a daydream that he can draw real and imaginary axes, with origin at the center of the table, in such a way that Tantrum is represented by a certain complex number z, and Overthetop is represented by the complex number $z + 1$. Breaking out of his daydream, Ivor suddenly exclaims "you are both sixth roots of 1!"

Prove that Ivor is correct, despite the incredulous editorial glares.

Chapter 7

Polynomial Equations

Expressions like $x^2 - 3x$, or $-7x^{102} + (3 - i)x^{17} - 7$, or more generally,

$$p(x) = a_n x^n + a_{n-1} x^{n-1} + \cdots + a_1 x + a_0 \, ,$$

where the coefficients $a_0, a_1, \ldots, a_n$ are complex numbers, are called *polynomials* in x. A *polynomial equation* is an equation of the form

$$p(x) = 0$$

where $p(x)$ is a polynomial. The *degree* of such an equation is the highest power of x which appears with a non-zero coefficient.

For example, equations of degree 1 take the form $ax + b = 0$, and are also known as *linear* equations; degree 2 equations $ax^2 + bx + c = 0$ are *quadratic* equations; degree 3 equations are *cubic* equations; degree 4 are *quartic* equations, degree 5 are *quintic* equations, and so on.

A complex number α is said to be a *root* of the polynomial equation $p(x) = 0$ if $p(\alpha) = 0$: in other words, when α is substituted for x, $p(x)$ becomes equal to 0. For example, 1 is a root of the cubic equation $x^3 - 3x + 2 = 0$.

The search for formulae for the roots of polynomial equations was one of the driving forces in mathematics from the time of the Greeks until the 19th century. Let us now taste a tiny flavor of this huge subject, in the hope that appetites are whetted for more.

It is obvious that any linear equation $ax + b = 0$ has exactly one root, namely $-\frac{b}{a}$. We are also familiar with the fact that any quadratic equation $ax^2 + bx + c = 0$ has roots in $\mathbb{C}$, given by the formula $\frac{1}{2a}(-b \pm \sqrt{b^2 - 4ac})$.

Things are less clear for cubic equations. Indeed, while the formula for the roots of a quadratic was known to the Greeks, it was not until the 16th century that a method for finding the roots of a cubic was found by the Italian mathematicians Scipio Ferreo, Tartaglia and Cardan. Here is their method.

Solution of Cubic Equations

Consider the cubic equation

$$x^3 + ax^2 + bx + c = 0 \tag{7.1}$$

The first step is to get rid of the x^2 term. This is easily done: put $y = x + \frac{a}{3}$. Then $y^3 = (x + \frac{a}{3})^3 = x^3 + ax^2 + \frac{a^2}{3}x + \frac{a^3}{27}$, so Equation (7.1) becomes $y^3 + b'y + c' = 0$ for some b', c'. Write this equation as

$$y^3 + 3hy + k = 0 \tag{7.2}$$

(The coefficients $3h$ and k can easily be worked out, given a, b, c.)

Here comes the clever part. Write $y = u + v$. Then

$$y^3 = (u + v)^3 = u^3 + v^3 + 3u^2v + 3uv^2 = u^3 + v^3 + 3uv(u + v)$$
$$= u^3 + v^3 + 3uvy .$$

Hence, the cubic equation

$$y^3 - 3uvy - (u^3 + v^3) = 0 \tag{7.3}$$

has $u + v$ as a root.

Our aim now is to find u and v so that the coefficients in Equations (7.2) and (7.3) are matched up. To match the coefficients, we require

$$h = -uv, \quad k = -\left(u^3 + v^3\right) . \tag{7.4}$$

From the first of these equations we have $v^3 = \frac{-h^3}{u^3}$, hence the second equation gives $u^3 - \frac{h^3}{u^3} = -k$, so

$$u^6 + ku^3 - h^3 = 0 . \tag{7.5}$$

This is just a quadratic equation for u^3, and a solution is

$$u^3 = \frac{1}{2}\left(-k + \sqrt{k^2 + 4h^3}\right) .$$

Then from (7.4),

$$v^3 = -k - u^3 = \frac{1}{2}\left(-k - \sqrt{k^2 + 4h^3}\right) .$$

As $y = u + v$, we have obtained the following formula for the roots of the cubic (7.2):

$$\sqrt[3]{\frac{1}{2}\left(-k + \sqrt{k^2 + 4h^3}\right)} + \sqrt[3]{\frac{1}{2}\left(-k - \sqrt{k^2 + 4h^3}\right)} .$$

Since a complex number has three cube roots, and there are two cube roots to be chosen, it seems that there are nine possible values for this formula. However, the equation $uv = -h$ implies that $v = \frac{-h}{u}$, and hence there are only three roots of (7.2), these being $u - \frac{h}{u}$ for each of the three choices for u.

Specifically, if u is one of the cube roots of $\frac{1}{2}(-k+\sqrt{k^2 + 4h^3})$, the other cube roots are $u\omega$, $u\omega^2$ where $\omega = e^{\frac{2\pi i}{3}}$, and so the roots of the cubic equation (7.2) are

$$ u - \frac{h}{u}, \quad u\omega - \frac{h\omega^2}{u}, \quad u\omega^2 - \frac{h\omega}{u} . $$

Once we know the roots of (7.2), we can of course write down the roots of the general cubic (7.1), since $x = y - \frac{a}{3}$.

Let us illustrate this method with a couple of examples.

Examples

(1) Consider the cubic equation $x^3 - 6x - 9 = 0$. This is (7.2) with $h = -2, k = -9$, so $\frac{1}{2}(-k+\sqrt{k^2 + 4h^3}) = \frac{1}{2}(9+\sqrt{49}) = 8$. Hence taking $u = 2$, we see that the roots are $3, 2\omega + \omega^2, 2\omega^2 + \omega$. As $\omega = \frac{1}{2}(-1 + i\sqrt{3})$ and $\omega^2 = \frac{1}{2}(-1 - i\sqrt{3})$, these roots are

$$ 3, \quad \frac{1}{2}\left(-3 + i\sqrt{3}\right), \quad \frac{1}{2}\left(-3 - i\sqrt{3}\right) . $$

(Of course these could easily have been worked out by cleverly spotting that 3 is a root, and factorizing the equation as $(x - 3)(x^2 + 3x + 3) = 0$.)

(2) Consider the equation $x^3 - 6x - 40 = 0$. The above formula gives roots of the form

$$ \sqrt[3]{20 + 14\sqrt{2}} + \sqrt[3]{20 - 14\sqrt{2}} . $$

However, we cleverly also spot that 4 is a root. What is going on?

In fact, nothing very mysterious is going on. The real cube root of $20 \pm 14\sqrt{2}$ is $2 \pm \sqrt{2}$, as can be seen by cubing the latter. Hence, the roots of the cubic are

$$ \left(2 + \sqrt{2}\right) + \left(2 - \sqrt{2}\right) = 4 , $$
$$ \left(2 + \sqrt{2}\right)\omega + \left(2 - \sqrt{2}\right)\omega^2 = -2 + i\sqrt{6} , $$
$$ \left(2 + \sqrt{2}\right)\omega^2 + \left(2 - \sqrt{2}\right)\omega = -2 - i\sqrt{6} . $$

Higher Degrees

Not long after the solution of the cubic, Ferrari, a pupil of Cardan, showed how to obtain a formula for the roots of a general quartic (degree 4) equation. The next step, naturally enough, was the quintic. However, several hundred years passed without anyone finding a formula for the roots of a general quintic equation.

There was a good reason for this. There is no such formula. Nor is there a formula for equations of degree greater than 5. This amazing fact was first established in the early 19th century by the Danish mathematician Abel (who died at age 26), after which the Frenchman Galois (who died at age 21) built an entirely new theory of equations, linking them to the then recent subject of group theory, which not only explained the non-existence of formulae, but laid the foundations of a whole edifice of algebra and number theory known as *Galois theory,* a major area of modern day research. If you get a chance, take a course in Galois theory during the rest of your studies in mathematics — you won't regret it!

The Fundamental Theorem of Algebra

So, there is no formula for the roots of a polynomial equation of degree 5 or more. We are therefore led to the troubling question: Can we be sure that such an equation actually has a root in the complex numbers?

The answer to this is yes, we can be sure. This is a famous theorem of another great mathematician — perhaps the greatest of all — Gauss:

THEOREM 7.1 *Fundamental Theorem of Algebra*
Every polynomial equation of degree at least 1 *has a root in* $\mathbb{C}$.

This is really a rather amazing result. After all, we introduced complex numbers just to be able to talk about roots of quadratics like $x^2 + 1 = 0$, and we find ourselves with a system which contains roots of *all* polynomial equations.

There are many different proofs of the Fundamental Theorem of Algebra available — Gauss himself found five during his lifetime. Probably the proofs that are easiest to understand are those using various basic results in the subject of Complex Analysis, and most undergraduate courses on this topic would include a proof of the Fundamental Theorem of Algebra.

The Fundamental Theorem of Algebra has many consequences. We shall just give a couple here. Let $p(x)$ be a polynomial of degree n. By the Fundamental Theorem 7.1, $p(x)$ has a root in $\mathbb{C}$, say α_1. It is rather easy to see from this (but we won't prove it here) that there is another polynomial $q(x)$, of degree $n - 1$, such that

$$p(x) = (x - \alpha_1) q(x) .$$

By Theorem 7.1, $q(x)$ also has a root in $\mathbb{C}$, say α_2, so as above there is a polynomial $r(x)$ of degree $n - 2$ such that

$$p(x) = (x - \alpha_1) (x - \alpha_2) r(x) .$$

Repeat this argument until we get down to a polynomial of degree 1. We thus obtain a factorization

$$p(x) = a (x - \alpha_1) (x - \alpha_2) \ldots (x - \alpha_n) ,$$

where a is the coefficient of x^n, and $\alpha_1, \ldots, \alpha_n$ are the roots of $p(x)$. These may of course not all be different, but there are precisely n of them if we count repeats.

Summarizing:

THEOREM 7.2

Every polynomial of degree n factorizes as a product of linear polynomials, and has exactly n roots in $\mathbb{C}$ (counting repeats).

Next, we briefly consider *real* polynomial equations — that is, equations of the form

$$p(x) = a_n x^n + a_{n-1} x^{n-1} + \cdots + a_1 x + a_0 = 0 ,$$

where all the coefficients $a_0, a_1, \ldots, a_n$ are real numbers.

Of course not all the roots of such an equation need be real (think of $x^2 + 1 = 0$). However, we can say something interesting about the roots.

Let $\alpha \in \mathbb{C}$ be a root of the real polynomial equation $p(x) = 0$. Thus,

$$p(\alpha) = a_n \alpha^n + a_{n-1} \alpha^{n-1} + \cdots + a_1 \alpha + a_0 = 0 .$$

Consider the complex conjugate $\bar{\alpha}$. We shall show this is also a root. To see this, observe first that $\overline{\alpha^2} = \bar{\alpha}^2$ (just apply Exercise 1(b) of Chapter 6 with $u = v = \alpha$); likewise, $\overline{\alpha^3} = \bar{\alpha}^3, \ldots, \overline{\alpha^n} = \bar{\alpha}^n$. Also $\bar{a}_i = a_i$ for all i, since the a_i are all real. Consequently,

$$
\begin{aligned}
p(\bar{\alpha}) &= a_n \bar{\alpha}^n + a_{n-1} \bar{\alpha}^{n-1} + \cdots + a_1 \bar{\alpha} + a_0 \\
&= a_n \overline{\alpha^n} + a_{n-1} \overline{\alpha^{n-1}} + \cdots + a_1 \bar{\alpha} + a_0 \\
&= \overline{a_n \alpha^n} + \overline{a_{n-1} \alpha^{n-1}} + \cdots + \overline{a_1 \alpha} + \overline{a_0} \\
&= \overline{p(\alpha)} = 0 ,
\end{aligned}
$$

showing that $\bar{\alpha}$ is indeed a root.

Thus for a real polynomial equation $p(x) = 0$, the non-real roots appear in complex conjugate pairs $\alpha, \bar{\alpha}$. Say the real roots are $\beta_1, \ldots, \beta_k$ and the non-real roots are $\alpha_1, \bar{\alpha}_1, \ldots, \alpha_l, \bar{\alpha}_l$ (where $k + 2l = n$). Then, as discussed above,

$$p(x) = (x - \beta_1) \ldots (x - \beta_k)(x - \alpha_1)(x - \bar{\alpha}_1) \ldots (x - \alpha_l)(x - \bar{\alpha}_l) .$$

Notice that $(x - \alpha_i)(x - \bar{\alpha}_i) = x^2 - (\alpha_i + \bar{\alpha}_i)x + \alpha_i\bar{\alpha}_i$, which is a quadratic with real coefficients. Thus, $p(x)$ factorizes as a product of real linear and real quadratic polynomials.

Summarizing:

THEOREM 7.3

Every real polynomial factorizes as a product of real linear and real quadratic polynomials, and has its non-real roots appearing in complex conjugate pairs.

Relationships Between Roots

Despite the fact that there is no general formula for the roots of a polynomial equation of degree 5 or more, there are still some interesting and useful relationships between the roots and the coefficients.

First consider a quadratic equation $x^2 + ax + b = 0$. If α, β are the roots, then $x^2 + ax + b = (x - \alpha)(x - \beta) = x^2 - (\alpha + \beta)x + \alpha\beta$, and hence, equating coefficients, we have

$$\alpha + \beta = -a, \quad \alpha\beta = b .$$

Likewise, for a cubic equation $x^3 + ax^2 + bx + c = 0$ with roots α, β, γ, we have $x^3 + ax^2 + bx + c = (x - \alpha)(x - \beta)(x - \gamma)$, hence

$$\alpha + \beta + \gamma = -a, \quad \alpha\beta + \alpha\gamma + \beta\gamma = b, \quad \alpha\beta\gamma = -c .$$

Applying this argument to an equation of degree n, we have

PROPOSITION 7.1

Let the roots of the equation

$$x^n + a_{n-1}x^{n-1} + \cdots + a_1x + a_0 = 0$$

be $\alpha_1, \alpha_2, \ldots, \alpha_n$. If s_1 denotes the sum of the roots, and s_2 denotes the sum of all products of pairs of roots, and s_3 denotes the sum of all products of triples *of roots, and so on, then*

$$s_1 = \alpha_1 + \cdots + \alpha_n = -a_{n-1} ,$$
$$s_2 = a_{n-2} ,$$
$$s_3 = -a_{n-3} ,$$
$$\cdots \quad \cdots$$
$$s_n = \alpha_1 \alpha_2 \ldots \alpha_n = (-1)^n a_0 .$$

PROOF We have

$$x^n + a_{n-1} x^{n-1} + \cdots + a_1 x + a_0 = (x - \alpha_1)(x - \alpha_2) \ldots (x - \alpha_n) .$$

If we multiply out the right-hand side, the coefficient of x^{n-1} is $-(\alpha_1 + \cdots + \alpha_n) = -s_1$, the coefficient of x^{n-2} is s_2, and so on. The result follows. ∎

Here are some examples of what can be done with this result.

Examples

(1) Write down a cubic equation with roots $1 + i$, $1 - i$, 2.

Answer If we call these three roots $\alpha_1, \alpha_2, \alpha_3$, then $s_1 = \alpha_1 + \alpha_2 + \alpha_3 = 4$, $s_2 = \alpha_1 \alpha_2 + \alpha_1 \alpha_3 + \alpha_2 \alpha_3 = (1 + i)(1 - i) + 2(1 + i) + 2(1 - i) = 6$ and $s_3 = \alpha_1 \alpha_2 \alpha_3 = 4$. Hence, a cubic with these roots is

$$x^3 - 4x^2 + 6x - 4 = 0 .$$

(2) If α, β are the roots of the equation $x^2 - 5x + 9 = 0$, find a quadratic equation with roots α^2, β^2.

Answer Of course this could be done by using the formula to write down α and β, then squaring them, and so on; this would be rather tedious, and a much more elegant solution is as follows. From the above, we have

$$\alpha + \beta = 5, \quad \alpha\beta = 9 .$$

Therefore, $\alpha^2 + \beta^2 = (\alpha + \beta)^2 - 2\alpha\beta = 5^2 - 18 = 7$, and $\alpha^2 \beta^2 = 9^2 = 81$. We therefore know the sum and product of α^2 and β^2, so a quadratic having these as roots is $x^2 - 7x + 81 = 0$.

(3) Find the value of k if the roots of the cubic equation $x^3 + x^2 + 2x + k = 0$ are in geometric progression.

Answer Saying that the roots are in geometric progression means that they are of the form α, αr, αr^2 for some r. We then have

$$\alpha \left(1 + r + r^2\right) = -1, \quad \alpha^2 \left(r + r^2 + r^3\right) = 2, \quad \alpha^3 r^3 = -k \,.$$

Dividing the first two of these gives $\alpha r = -2$. Hence, the third gives $k = 8$.

Exercises for Chapter 7

1. Use the method given above for solving cubics to find the roots of the equation $x^3 - 6x^2 + 13x - 12 = 0$.

   Now notice that 3 is one of the roots. Reconcile this with the roots you have found.

2. Solve $x^3 - 15x - 4 = 0$ using the method for solving cubics.

   Now cleverly spot an integer root. Deduce that

   $$\cos\left(\frac{1}{3}\tan^{-1}\left(\frac{11}{2}\right)\right) = \frac{2}{\sqrt{5}}.$$

3. Factorize $x^5 + 1$ as a product of real linear and quadratic polynomials.

   Do the same with $x^n - 1$ (where n is a positive integer).

4. (a) Find the value of k such that the roots of $x^3 + 6x^2 + kx - 10 = 0$ are in arithmetical progression (i.e., are α, $\alpha + d$, $\alpha + 2d$ for some d). Solve the equation for this value of k.

   (b) If the roots of the equation $x^3 - x - 1 = 0$ are α, β, γ, find a cubic equation having roots α^2, β^2, γ^2, and also a cubic equation having roots $\frac{1}{\alpha}$, $\frac{1}{\beta}$, $\frac{1}{\gamma}$.

   (c) Given that the sum of two of the roots of the equation $x^3 + px^2 + p^2x + r = 0$ is 1, prove that $r = (p+1)(p^2 + p + 1)$.

   (b) Solve the equation $x^4 - 3x^3 - 5x^2 + 17x - 6 = 0$, given that the sum of two of the roots is 5.

5. Find all real or complex solutions of the simultaneous equations

   $$x + y + z = 3$$
   $$x^2 + y^2 + z^2 = 3$$
   $$x^3 + y^3 + z^3 = 3$$

   (*Hint*: Look for a cubic equation which has roots x, y, z.)

Chapter 8

Induction

Consider the following three statements, each involving a general positive integer n:

(1) The sum of the first n odd numbers is equal to n^2.

(2) If $p > -1$ then $(1 + p)^n \geq 1 + np$.

(3) The sum of the internal angles in an n-sided polygon is $(n - 2)\pi$.

(A *polygon* is a closed figure with straight edges, such as a triangle (3 sides), a quadrilateral (4 sides), a pentagon (5 sides), etc.)

We can check that these statements are true for various specific values of n. For instance, (1) is true for $n = 2$ as $1 + 3 = 4 = 2^2$, and for $n = 3$ as $1 + 3 + 5 = 9 = 3^2$; statement (2) is true for $n = 1$ as $1 + p \geq 1 + p$, and for $n = 2$ as $(1 + p)^2 = 1 + 2p + p^2 \geq 1 + 2p$; and (3) is true for $n = 3$ as the sum of the angles in a triangle is π, and for $n = 4$ as the sum of the angles in a quadrilateral is 2π.

But how do we go about trying to prove the truth of these statements for *all* values of n?

The answer is that we use the following basic principle. In it we denote by $P(n)$ a statement involving a positive integer n; for example, $P(n)$ could be any of statements (1), (2), or (3) above.

Principle of Mathematical Induction
Suppose that for each positive integer n we have a statement $P(n)$. If we prove the following two things:
  (a) $P(1)$ *is true;*
  (b) *for all n, if $P(n)$ is true then $P(n + 1)$ is also true;*
then $P(n)$ is true for all positive integers n.

The logic behind this principle is clear: by (a), the first statement $P(1)$ is true. By (b) with $n = 1$, we know that $P(1) \Rightarrow P(2)$, hence $P(2)$ is true. By (b) with $n = 2$, $P(2) \Rightarrow P(3)$, hence $P(3)$ is true; and so on.

The principle may look a little strange at first sight, but a few examples should clarify matters.

Example 8.1

Let us try to prove statement (1) above using the Principle of Mathematical Induction. Here $P(n)$ is the statement that the sum of the first n odd numbers is n^2. In other words:

$$P(n) : 1 + 3 + 5 + \cdots + 2n - 1 = n^2 .$$

We need to carry out parts (a) and (b) of the Principle.

(a) $P(1)$ is true, since $1 = 1^2$.

(b) Suppose $P(n)$ is true. Then

$$1 + 3 + 5 + \cdots + 2n - 1 = n^2 .$$

Adding $2n + 1$ to both sides gives

$$1 + 3 + 5 + \cdots + 2n - 1 + 2n + 1 = n^2 + 2n + 1 = (n + 1)^2 ,$$

which is statement $P(n + 1)$. Thus, we have shown that $P(n) \Rightarrow P(n + 1)$.

We have now established parts (a) and (b). Hence by the Principle of Mathematical Induction, $P(n)$ is true for all positive integers n.

The phrase "Principle of Mathematical Induction" is quite a mouthful, and we usually use just the single word "induction" instead.

Example 8.2

Now let us prove statement (2) above by induction. Here, for n a positive integer $P(n)$ is the statement

$$P(n) : \text{ if } p > -1 \text{ then } (1 + p)^n \geq 1 + np .$$

For (a), observe $P(1)$ is true, as $1 + p \geq 1 + p$.

For (b), suppose $P(n)$ is true, so $(1 + p)^n \geq 1 + np$. Since $p > -1$ we know that $1 + p > 0$, so we can multiply both sides of the inequality by $1 + p$ (see Example 4.3) to obtain

$$(1 + p)^{n+1} \geq (1 + np)(1 + p) = 1 + (n + 1)p + np^2 .$$

Since $np^2 \geq 0$, this implies that $(1 + p)^{n+1} \geq 1 + (n + 1)p$, which is statement $P(n + 1)$. Thus we have shown $P(n) \Rightarrow P(n + 1)$.

Therefore, by induction, $P(n)$ is true for all positive integers n.

Next we attempt to prove the statement (3) concerning n-sided polygons. There is a slight problem here. If we naturally enough let $P(n)$ be statement (3), then $P(n)$ only makes sense if $n \geq 3$; $P(1)$ and $P(2)$ make no sense, as there is no such thing as a 1-sided or 2-sided polygon. To take care of such a situation, we need a slightly modified Principle of Mathematical Induction:

Principle of Mathematical Induction II

Let k be an integer. Suppose that for each integer $n \geq k$ we have a statement $P(n)$. If we prove the following two things:

(a) *$P(k)$ is true;*

(b) *for all $n \geq k$, if $P(n)$ is true then $P(n + 1)$ is also true;*

then $P(n)$ is true for all integers $n \geq k$.

The logic behind this is the same as explained before.

Example 8.3

Now we prove statement (3). Here we have $k = 3$ in the above Principle, and for $n \geq 3$, $P(n)$ is the statement

$P(n)$: the sum of the internal angles in an n-sided polygon is $(n - 2)\pi$.

For (a), observe that $P(3)$ is true, since the sum of the angles in a triangle is $\pi = (3 - 2)\pi$.

Now for (b). Suppose $P(n)$ is true. Consider an $(n + 1)$-sided polygon with corners $A_1, A_2, \ldots, A_{n+1}$:

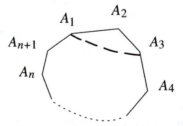

Draw the line $A_1 A_3$. Then $A_1 A_3 A_4 \ldots A_{n+1}$ is an n-sided polygon. Since we are assuming $P(n)$ is true, the internal angles in this n-sided polygon add up to $(n - 2)\pi$. From the picture we see that the sum of the angles in the $(n + 1)$-sided polygon $A_1 A_2 \ldots A_{n+1}$ is equal to the sum of those in $A_1 A_3 A_4 \ldots A_{n+1}$, plus the sum of those in the triangle $A_1 A_2 A_3$, hence is

$$(n - 2)\pi + \pi = ((n + 1) - 2)\pi .$$

We have now shown that $P(n) \Rightarrow P(n + 1)$. Hence by induction, $P(n)$ is true for all $n \geq 3$.

The next example also uses the slightly modified Principle of Mathematical Induction II. In it, for a positive integer n we define

$$n! = n(n-1)(n-2)\ldots 3\cdot 2\cdot 1 \, ,$$

the product of all the integers between 1 and n. The symbol $n!$ is usually referred to as n *factorial*. By convention, we also define $0! = 1$.

Example 8.4

For which positive integers n is $2^n < n!$?

Answer Let $P(n)$ be the statement that $2^n < n!$. Observe that

$$2^1 > 1!, \quad 2^2 > 2!, \quad 2^3 > 3!, \quad 2^4 < 4!, \quad 2^5 < 5! \, ,$$

so $P(1)$, $P(2)$, $P(3)$ are false, while $P(4)$, $P(5)$ are true. Therefore, it seems sensible to try to prove $P(n)$ is true for all $n \geq 4$.

First, $P(4)$ is true, as observed above.

Now suppose n is an integer at least 4, and $P(n)$ is true. Thus

$$2^n < n!$$

Multiplying both sides by 2, we get

$$2^{n+1} < 2(n!) \, .$$

Since $2 < n+1$, $2(n!) < (n+1)n! = (n+1)!$, and hence $2^{n+1} < (n+1)!$. This shows that $P(n) \Rightarrow P(n+1)$. Therefore, by induction, $P(n)$ is true for all $n \geq 4$.

Guessing the Answer

Some problems cannot immediately be tackled using induction, but first require some intelligent guesswork. Here is an example.

Example 8.5

Find a formula for the sum

$$\frac{1}{1\cdot 2} + \frac{1}{2\cdot 3} + \cdots + \frac{1}{n(n+1)} \, .$$

Answer Calculate this sum for the first few values of n:

$$n = 1 : \frac{1}{1 \cdot 2} = \frac{1}{2},$$

$$n = 2 : \frac{1}{1 \cdot 2} + \frac{1}{2 \cdot 3} = \frac{1}{2} + \frac{1}{6} = \frac{2}{3},$$

$$n = 3 : \frac{1}{1 \cdot 2} + \frac{1}{2 \cdot 3} + \frac{1}{3 \cdot 4} = \frac{3}{4}.$$

We intelligently spot a pattern in these answers, and guess that the sum of n terms is probably $\frac{n}{n+1}$. Hence we let $P(n)$ be the statement

$$P(n) : \frac{1}{1 \cdot 2} + \frac{1}{2 \cdot 3} + \cdots + \frac{1}{n(n+1)} = \frac{n}{n+1}$$

and attempt to prove $P(n)$ true for all $n \geq 1$ by induction.

First, $P(1)$ is true, as noted above.

Now assume $P(n)$ is true, so

$$\frac{1}{1 \cdot 2} + \frac{1}{2 \cdot 3} + \cdots + \frac{1}{n(n+1)} = \frac{n}{n+1}.$$

Adding $\frac{1}{(n+1)(n+2)}$ to both sides gives

$$\frac{1}{1 \cdot 2} + \cdots + \frac{1}{n(n+1)} + \frac{1}{(n+1)(n+2)} = \frac{n}{n+1} + \frac{1}{(n+1)(n+2)}$$

$$= \frac{n(n+2) + 1}{(n+1)(n+2)} = \frac{n^2 + 2n + 1}{(n+1)(n+2)} = \frac{(n+1)^2}{(n+1)(n+2)} = \frac{n+1}{n+2}.$$

Hence $P(n) \Rightarrow P(n+1)$. So by induction $P(n)$ is true for all $n \geq 1$.

The Σ Notation

Before proceeding with the next example, we introduce an important notation for writing down sums of many terms. If $f_1, f_2, \ldots, f_n$ are numbers, we abbreviate the sum of all of them by

$$f_1 + f_2 + \cdots + f_n = \sum_{r=1}^{n} f_r .$$

(The symbol Σ is the Greek capital letter "Sigma," so this is often called the "Sigma notation.") For example, setting $f_r = \frac{1}{r(r+1)}$, we have

$$\frac{1}{1 \cdot 2} + \frac{1}{2 \cdot 3} + \cdots + \frac{1}{n(n+1)} = \sum_{r=1}^{n} \frac{1}{r(r+1)} .$$

Thus, Example 8.5 says that

$$\sum_{r=1}^{n} \frac{1}{r(r+1)} = \frac{n}{n+1} ,$$

and Example 8.1 says

$$\sum_{r=1}^{n} (2r-1) = n^2 .$$

Notice that if a, b, c are constants, then

$$\sum_{r=1}^{n} (af_r + bg_r + c) = a \sum_{r=1}^{n} f_r + b \sum_{r=1}^{n} g_r + cn , \qquad (*)$$

since the left-hand side is equal to

$$(af_1 + bg_1 + c) + \cdots + (af_n + bg_n + c)$$
$$= a(f_1 + \cdots + f_n) + b(g_1 + \cdots + g_n) + (c + \cdots + c) ,$$

which is the right-hand side.

The Equation $(*)$ is quite useful for manipulating sums. Here is an elementary example using it.

Example 8.6

Find a formula for $\sum_{r=1}^{n} r \ (= 1 + 2 + \cdots + n).$

Answer Write $s_n = \sum_{r=1}^{n} r$. By Example 8.1, $\sum_{r=1}^{n} (2r-1) = n^2$, so using Equation $(*)$,

$$n^2 = \sum_{r=1}^{n} (2r-1) = 2 \sum_{r=1}^{n} r - n = 2s_n - n .$$

Hence, $s_n = \frac{1}{2}n(n+1)$.

So we know the sum of the first n positive integers. What about the sum of the first n squares?

Example 8.7

Find a formula for $\sum_{r=1}^{n} r^2 \ (= 1^2 + 2^2 + \cdots + n^2).$

Answer We first try to guess the answer (intelligently). The first few values $n = 1, 2, 3, 4$ give sums 1,5,14,30. It is not easy to guess a formula from these values, so yet a smidgeon more intelligence is required. The sum we are trying to find is the sum of n terms of a quadratic nature, so it seems reasonable to look for a formula for the sum which is a cubic in n, say $an^3 + bn^2 + cn + d$.

What should a, b, c, d be? Well, they have to fit in with the values of the sum for $n = 1, 2, 3, 4$, and hence must satisfy the following equations:

$$n = 1 : 1 = a + b + c + d \tag{8.1}$$
$$n = 2 : 5 = 8a + 4b + 2c + d \tag{8.2}$$
$$n = 3 : 14 = 27a + 9b + 3c + d \tag{8.3}$$
$$n = 4 : 30 = 64a + 16b + 4c + d \tag{8.4}$$

Equations (8.2)-(8.1), (8.3)-(8.2), (8.4)-(8.3) then give $4 = 7a + 3b + c$, $9 = 19a+5b+c$, $16 = 37a+7b+c$. Subtraction of these gives $5 = 12a+2b$, $7 = 18a + 2b$. Hence we get the solution

$$a = \frac{1}{3}, \ b = \frac{1}{2}, \ c = \frac{1}{6}, \ d = 0 .$$

Hence, our (intelligent) guess is that

$$\sum_{r=1}^{n} r^2 = \frac{1}{3}n^3 + \frac{1}{2}n^2 + \frac{1}{6}n = \frac{1}{6}n(n + 1)(2n + 1) .$$

This turns out to be correct, and we leave it to the reader to prove it by induction. (It is set as Exercise 1 at the end of the chapter in case you forget.)

(Actually, there is a much better way of working out a formula for $\sum_{r=1}^{n} r^2$, given in Exercise 3 at the end of the chapter.)

The next example is a nice geometric proof by induction.

Example 8.8

Lines in the plane. If we draw a straight line in the plane, it divides the plane into two regions. If we draw another, not parallel to the first, the two lines divide the plane into four regions. Likewise, three lines, not all going through the same point, and no two of which are parallel, divide the plane into seven regions:

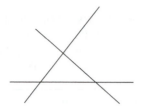

We can carry on drawing lines and counting the regions they form, which leads us naturally to a general question:

If we draw n straight lines in the plane, no three going through the same point, and no two parallel, how many regions do they divide the plane into?

The conditions about not going through the same point and not being parallel may seem strange, but in fact they are very natural: if you draw lines at random, it is very unlikely that two will be parallel or that three will pass through the same point — so you could say the lines in the question are "random" lines. Technically, they are said to be *lines in general position*.

The answers to the question for $n = 1, 2, 3, 4$ are 2, 4, 7, 11. Even from this flimsy evidence you have probably spotted a pattern — the difference between successive terms seems to be increasing by 1 each time. Can we predict a formula from this pattern? Yes, of course we can: the number of regions for one line is two, for two lines is $2 + 2$, for three lines is $2 + 2 + 3$, for four lines is $2 + 2 + 3 + 4$, so we predict that the number of regions for n lines is

$$2 + 2 + 3 + 4 + \cdots + n .$$

This is just $1 + \Sigma_{r=1}^{n} r$, which by Example 8.6 is equal to $1 + \frac{1}{2}n(n + 1)$.

Let us therefore attempt to prove the following statement $P(n)$ by induction: the number of regions formed in the plane by n straight lines in general position is $\frac{1}{2}(n^2 + n + 2)$.

First $P(1)$ is true, as the number of regions for one line is $2 = \frac{1}{2}(1^2 + 1 + 2)$.

Now suppose $P(n)$ is true, so n lines in general position form $\frac{1}{2}(n^2 + n + 2)$ regions. Draw in an $(n+1)^{th}$ line. Since it is not parallel to any of the others, this line meets each of the other n lines in a point, and these n points of intersection divide the $(n + 1)^{th}$ line into $n + 1$ pieces. Each of these pieces divides an old region into two new ones. Hence, when the $(n + 1)^{th}$ line is drawn, the number of regions increases by $n + 1$. (If this argument is not clear to you, try drawing a picture to illustrate it when $n = 3$ or 4.) Consequently the number of regions with $n + 1$ lines is equal to $\frac{1}{2}(n^2 + n + 2) + n + 1$. Check that this is equal to $\frac{1}{2}((n + 1)^2 + (n + 1) + 2)$.

We have now shown that $P(n) \Rightarrow P(n + 1)$. Hence, by induction, $P(n)$ is true for all $n \geq 1$.

Induction is a much more powerful method than you might think. It can often be used to prove statements that do not actually explicitly mention an integer n.

In such instances, one must imaginatively design a suitable statement $P(n)$ to fit in with the problem, and then try to prove $P(n)$ by induction. In the next two examples this is fairly easy to do. The next chapter, however, will be devoted to an example of a proof by induction where the statement $P(n)$ lies a long way away from the initial problem.

Example 8.9

Some straight lines are drawn in the plane, forming regions as in the previous example. Show that it is possible to color each region either red or blue, in such a way that no two neighboring regions have the same color.

For example, here is such a coloring when there are three lines:

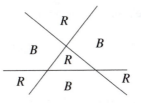

How do we design a suitable statement $P(n)$ for this problem? This is very simple: just take $P(n)$ to be the statement that the regions formed by n straight lines and the plane can be colored in the required way.

Actually, the proof of $P(n)$ by induction is so neat and elegant that I would hate to deprive you of the pleasure of thinking about it, so I leave it to you. (It is set as Exercise 11 at the end of the chapter in case you forget.)

Prime Factorization

In the next example, we prove a very important result about the integers. First we need a definition:

DEFINITION   A prime number *is a positive integer p such that $p \geq 2$ and the only positive integers dividing p are 1 and p.*

You are probably familiar to some extent with prime numbers. The first few are 2, 3, 5, 7, 11, 13, 17, 19, 23, 29, ...

The important result we shall prove is the following:

PROPOSITION 8.1

Every positive integer greater than 1 is equal to a product of prime numbers.

In the proposition, the number of primes in a product must be allowed to be 1, since a prime number itself is a product of one prime. If n is a positive integer, we call an expression $n = p_1 \ldots p_k$, where $p_1, \ldots, p_k$ are prime numbers, a *prime factorization* of n. Here are some examples of prime factorizations:

$$30 = 2 \cdot 3 \cdot 5, \quad 12 = 2 \cdot 2 \cdot 3, \quad 13 = 13 \, .$$

A suitable statement to attempt to prove by induction is easy to design: for $n \geq 2$, let $P(n)$ be the statement that n is equal to a product of prime numbers.

Clearly $P(2)$ is true, as $2 = 2$ is a prime factorization of 2. However, it is not clear at all how to go about showing that $P(n) \Rightarrow P(n + 1)$. In fact this cannot be done, since the primes in the prime factorization of n do not occur in the factorization of $n + 1$.

However, all is not lost. We shall use the following, apparently stronger, principle of induction.

Principle of Strong Mathematical Induction

Suppose that for each integer $n \geq k$ we have a statement $P(n)$. If we prove the following two things:

(a) $P(k)$ is true;

(b) for all n, if $P(k), P(k + 1), \ldots, P(n)$ are all true, then $P(n + 1)$ is also true;

then $P(n)$ is true for all $n \geq k$.

The logic behind this principle is not really any different from that behind the old principle: by (a), $P(k)$ is true. By (b), $P(k) \Rightarrow P(k + 1)$, hence $P(k + 1)$ is true. By (b) again, $P(k), P(k + 1) \Rightarrow P(k + 2)$, hence $P(k + 2)$ is true, and so on.

(In fact, the Principle of Strong Induction is actually implied by the old principle. To see this, let $Q(n)$ be the statement that all of $P(k), \ldots, P(n)$ are true. Suppose we have proved (a) and (b) of Strong Induction. Then by (a), $Q(k)$ is true, and by (b), $Q(n) \Rightarrow Q(n + 1)$. Hence by the old principle, $Q(n)$ is true for all $n \geq k$, and therefore so is $P(n)$.)

Let us now apply Strong Induction to prove Proposition 8.1.

Proof of Proposition 8.1   For $n \geq 2$, let $P(n)$ be the statement that n is equal to a product of prime numbers. As we have already remarked, $P(2)$ is true.

Now for part (b) of Strong Induction. Suppose that $P(2), \ldots, P(n)$ are all true. This means that every integer between 2 and n has a prime factorization.

Now consider $n+1$. If $n+1$ is prime, then it certainly has a prime factorization (as a product of 1 prime). If $n+1$ is not prime, then by the definition of a prime number, there is an integer a dividing $n+1$ such that $a \neq 1$ or $n+1$. Writing $b = \frac{n+1}{a}$, we then have

$$n + 1 = ab \quad \text{and} \quad a, b \in \{2, 3, \ldots, n\}.$$

By assumption, $P(a)$ and $P(b)$ are both true, i.e., a and b have prime factorizations. Say

$$a = p_1 \ldots p_k, \quad b = q_1 \ldots q_l,$$

where all the p_i and q_i are prime numbers. Then

$$n + 1 = ab = p_1 \ldots p_k q_1 \ldots q_l.$$

This is an expression for $n + 1$ as a product of prime numbers.

We have now shown that $P(2), \ldots, P(n) \Rightarrow P(n + 1)$. Therefore, $P(n)$ is true for all $n \geq 2$, by Strong Induction. ∎

Example 8.10

Suppose we are given a sequence of integers $u_0, u_1, u_2, \ldots, u_n, \ldots$ such that $u_0 = 2$, $u_1 = 3$ and

$$u_{n+1} = 3u_n - 2u_{n-1}$$

for all $n \geq 1$. (Such an equation is called a *recurrence relation* for the sequence.) Can we find a formula for u_n?

Using the equation with $n = 1$, we get $u_2 = 3u_1 - 2u_0 = 5$, and likewise $u_3 = 9, u_4 = 17, u_5 = 33, u_6 = 65$. Is there an obvious pattern? Yes, a reasonable guess seems to be that $u_n = 2^n + 1$.

So let us try to prove by Strong Induction that $u_n = 2^n + 1$. If this is the statement $P(n)$, then $P(0)$ is true, as $u_0 = 2^0 + 1 = 2$. Suppose $P(0), P(1), \ldots, P(n)$ are all true. Then $u_n = 2^n + 1$ and $u_{n-1} = 2^{n-1} + 1$. Hence from the recurrence relation,

$$u_{n+1} = 3\left(2^n + 1\right) - 2\left(2^{n-1} + 1\right) = 3 \cdot 2^n - 2^n + 1 = 2^{n+1} + 1,$$

which shows $P(n + 1)$ is true. Therefore, $u_n = 2^n + 1$ for all n, by Strong Induction.

Exercises for Chapter 8

1. Prove by induction that $\Sigma_{r=1}^{n} r^2 = \frac{1}{6}n(n+1)(2n+1)$.

   Deduce formulae for

   $$1.1 + 2.3 + 3.5 + 4.7 + \cdots + n(2n-1) \quad \text{and} \quad 1^2 + 3^2 + 5^2 + \cdots (2n-1)^2.$$

2. (a) Work out 1, $1+8$, $1+8+27$ and $1+8+27+64$. Guess a formula for $\Sigma_{r=1}^{n} r^3$ and prove it.

   (b) Check that $1 = 0+1$, $2+3+4 = 1+8$ and $5+6+\cdots+9 = 8+27$. Find a general formula for which these are the first three cases. Prove your formula is correct.

3. Here is another way to work out $\Sigma_{r=1}^{n} r^2$. Observe that $(r+1)^3 - r^3 = 3r^2 + 3r + 1$. Hence

   $$\sum_{r=1}^{n} (r+1)^3 - r^3 = 3\sum_{r=1}^{n} r^2 + 3\sum_{r=1}^{n} r + n.$$

   The left-hand side is equal to

   $$\left(2^3 - 1^3\right) + \left(3^3 - 2^3\right) + \left(4^3 - 3^3\right) + \cdots$$
   $$+ \left((n+1)^3 - n^3\right) = (n+1)^3 - 1.$$

   Hence we can work out $\Sigma_{r=1}^{n} r^2$.

   Carry out this calculation, and check that your formula agrees with that in Question 1.

   Use the same method to work out formulae for $\Sigma_{r=1}^{n} r^3$ and $\Sigma_{r=1}^{n} r^4$.

4. Prove the following statements by induction:

   (a) For all integers $n \geq 0$, the number $5^{2n} - 3^n$ is a multiple of 11.

   (b) The sum of the cubes of three consecutive positive integers is always a multiple of 9.

   (c) It is possible to pay, without requiring change, any whole number of roubles greater than 7, with banknotes of value 3 roubles and 5 roubles.

5. The *Fibonacci sequence* is a sequence of integers $u_1, u_2, \ldots, u_n, \ldots,$ such that $u_1 = 1$, $u_2 = 1$ and

$$u_{n+1} = u_n + u_{n-1}$$

for all $n \geq 1$. Prove by strong induction that for all n,

$$u_n = \frac{1}{\sqrt{5}} \left(\alpha^n - \beta^n \right),$$

where $\alpha = \frac{1+\sqrt{5}}{2}$ and $\beta = \frac{1-\sqrt{5}}{2}$.

6. Prove that if $0 < q < \frac{1}{2}$, then for all $n \geq 1$,

$$(1 + q)^n \leq 1 + 2^n q.$$

7. (a) Prove that for every integer $n \geq 2$,

$$\frac{1}{n+1} + \frac{1}{n+2} + \cdots + \frac{1}{2n} \geq \frac{7}{12}.$$

(b) Prove that for every integer $n \geq 1$,

$$1 + \frac{1}{\sqrt{2}} + \frac{1}{\sqrt{3}} + \cdots + \frac{1}{\sqrt{n}} < 2\sqrt{n}.$$

8. Just for this question, count 1 as a prime number. A well-known result in number theory says that for every integer $x \geq 3$, there is a prime number p such that $\frac{1}{2}x < p < x$. Using this result and strong induction, prove that every positive integer is equal to a sum of primes, all of which are different.

9. Here is a "proof" by induction that any two positive integers are equal (e.g., $5 = 10$):

First, a definition: if a and b are positive integers, define $\max(a, b)$ to be the larger of a and b if $a \neq b$, and to be a if $a = b$. (For instance, $\max(3, 5) = 5$, $\max(3, 3) = 3$.) Let $P(n)$ be the statement: "if a and b are positive integers such that $\max(a, b) = n$, then $a = b$." We prove $P(n)$ true for all $n \geq 1$ by induction. (As a consequence, if a, b are any two positive integers, then $a = b$, since $P(n)$ is true, where $n = \max(a, b)$.)

First, $P(1)$ is true, since if $\max(a, b) = 1$ then a and b must both be equal to 1. Now assume $P(n)$ true. Let a, b be positive integers such that $\max(a, b) = n + 1$. Then $\max(a - 1, b - 1) = n$. As we are assuming $P(n)$, this implies that $a - 1 = b - 1$, hence $a = b$. Therefore, $P(n + 1)$ is true. By induction, $P(n)$ is true for all n.

There must be something wrong with this "proof." Can you find the error?

10. (a) Suppose we have n straight lines in a plane, and all the lines pass through a single point. Into how many regions do the lines divide the plane? Prove your answer.

(b) We know from Example 8.8 that n straight lines in general position in a plane divide the plane into $\frac{1}{2}(n^2 + n + 2)$ regions. How many of these regions are infinite and how many are finite?

(In case of any confusion, a finite region is one that has finite area; an infinite region is one that does not.)

11. (See Example 8.9.) Some straight lines are drawn in the plane, forming regions. Show that it is possible to color each region either red or blue, in such a way that no two neighboring regions have the same color.

12. Critic Ivor Smallbrain is sitting through a showing of the new film *Polygon with the Wind*. Ivor is not enjoying the film, and begins to doodle on a piece of paper, drawing circles in such a way that any two of the circles intersect, no two circles touch each other, and no three circles pass through the same point. He notices that after drawing two circles he has divided the plane into four regions, after three there are eight regions, and wonders to himself how many regions there will be after he has drawn n circles. Can you help Ivor?

Chapter 9

Euler's Formula and Platonic Solids

This chapter contains a rather spectacular proof by induction. The result we shall prove is a famous formula of Euler from the 18th century, concerning the relationship between the numbers of corners, edges and faces of a solid object. As an application of Euler's formula we shall then study the five Platonic solids — the cube, regular tetrahedron, octahedron, icosahedron and dodecahedron.

We shall call our solid objects *polyhedra*. A *polyhedron* is a solid whose surface consists of a number of faces, all of which are polygons, such that any side of a face lies on exactly one other face. The corners of the faces are called the *vertices* of the polyhedron, and their sides are the *edges* of the polyhedron.

Here are some everyday examples of polyhedra.

(1) *Cube*

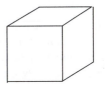

This has 8 vertices, 12 edges and 6 faces.

(2) *Tetrahedron*

This has 4 vertices, 6 edges and 4 faces.

(3) *Triangular prism*

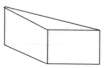

This has 6 vertices, 9 edges and 5 faces.

(4) *n-prism* This is like the triangular prism, except that its top and bottom faces are n-sided polygons rather than triangles. It has $2n$ vertices, $3n$ edges and $n + 2$ faces.

Let us collect the numbers of vertices, edges and faces for the above examples in a table. Denote these numbers by V, E and F, respectively.

	V	E	F
(1)	8	12	6
(2)	4	6	4
(3)	6	9	5
(4)	$2n$	$3n$	$n + 2$

Can you see a relationship between these numbers that holds in every case? You probably can — it is

$$V - E + F = 2.$$

This is Euler's famous formula, and we shall show that it holds in general for all *convex* polyhedra: a polyhedron is convex if, whenever we choose two points on its surface, the straight line joining them lies entirely within the polyhedron.

All of the above examples are convex polyhedra. However, if we for example take a cube and remove a smaller cube from its interior, we get a polyhedron which is not convex; for this polyhedron, in fact, $V = 16$, $E = 24$, $F = 12$, so $V - E + F = 4$ and the formula fails.

Here then is Euler's formula.

THEOREM 9.1

For a convex polyhedron with V vertices, E edges and F faces, we have

$$V - E + F = 2.$$

As I said, we shall prove this result by induction. So somehow we have to design a suitable statement $P(n)$ to try to prove by induction. What on earth should $P(n)$ be?

Before going into this, let us first translate the problem from one about objects in 3-dimensional space to one about objects in the plane. Take a convex polyhedron as in the theorem, and choose one face of it. Regard this face as a window, put your eye very close to the window, and draw on the window pane the vertices and edges you can see through the window. The result is a figure in the plane with straight edges, vertices and faces. For example, here is what we would draw for the cube, the tetrahedron and the triangular prism:

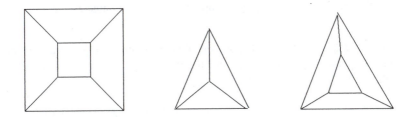

(The outer edges enclose the window.)

The resulting figure in the plane has V vertices, E edges and $F - 1$ faces (we lose one face, since the window is no longer a face). It is a "connected plane graph," in the sense of the following definition.

DEFINITION   *A plane graph is a figure in the plane consisting of a collection of points (vertices), and some edges joining various pairs of these points, with no two edges crossing each other. A plane graph is* connected *if we can get from any vertex of the graph to any other vertex by going along a path of edges in the graph.*

For example, here is a connected plane graph:

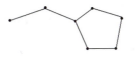

It has 7 vertices, 7 edges and 1 face.

THEOREM 9.2

If a connected plane graph has v vertices, e edges and f faces, then

$$v - e + f = 1 .$$

Theorem 9.2 easily implies Euler's theorem 9.1: for if we have a convex polyhedron with V vertices, E edges and F faces, then as explained above we get a connected plane graph with V vertices, E edges and $F - 1$ faces. If we knew Theorem 9.2 was true, we could then deduce that $V - E + (F - 1) = 1$, hence $V - E + F = 2$, as required for Euler's theorem.

So we need to prove Theorem 9.2.

Proof of Theorem 9.2   Here is the statement $P(n)$ that we are going to try to prove by induction:

$P(n)$: every connected plane graph with n edges satisfies the formula $v - n + f = 1$.

Notice that $P(n)$ is a statement about lots of plane graphs. $P(1)$ says that every connected plane graph with 1 edge satisfies the formula; there is only one such graph:

This has 2 vertices, 1 edge and 0 faces, and since $2 - 1 + 0 = 1$, it satisfies the formula. Likewise, $P(2)$ says that the graph

satisfies the formula, which it does, as $v = 3$, $e = 2$, $f = 0$. For $P(3)$, there are three different connected plane graphs with 3 edges:

Each satisfies the formula.

Let us prove $P(n)$ by induction. First, $P(1)$ is true, as observed in the previous paragraph.

Now assume $P(n)$ is true — so every connected plane graph with n edges satisfies the formula. We need to deduce $P(n + 1)$. So consider a connected plane graph G with $n + 1$ edges. Say G has v vertices and f faces. We want to prove that G satisfies the formula $v - (n + 1) + f = 1$.

Our strategy will be to remove a carefully chosen edge from G, so as to leave a connected plane graph with only n edges, and then use $P(n)$.

If G has at least 1 face (i.e., $f \geq 1$), we remove one edge of this face. The remaining graph G' is still connected, and has n edges, v vertices and $f - 1$ faces. Since we are assuming $P(n)$, we know that G' satisfies the formula, hence

$$v - n + (f - 1) = 1 .$$

Therefore $v - (n + 1) + f = 1$, as required.

If G has no faces at all (i.e., $f = 0$), then it has at least one "end-vertex," i.e., a vertex which is joined by an edge to only one other vertex. Removing this end-vertex and its edge from G leaves a connected plane graph G'' with $v - 1$ vertices, n edges and 0 faces. By $P(n)$, G'' satisfies the formula, so

$$(v - 1) - n + 0 = 1 .$$

Hence $v - (n + 1) + 0 = 1$, which is the formula for G.

We have established that $P(n) \Rightarrow P(n + 1)$, so $P(n)$ is true for all n by induction.

This completes the proof of Theorem 9.2, and hence of Euler's theorem 9.1.

Regular and Platonic Solids

A polygon is said to be *regular* if all its sides are of equal length and all its internal angles are equal. We call a regular polygon with n sides a *regular n-gon*. Some of these shapes are probably quite familiar; for example, a regular n-gon with $n = 3$ is just an equilateral triangle, $n = 4$ is a square, $n = 5$ is a regular pentagon, and so on:

A polyhedron is *regular* if its faces are regular polygons, all with the same number of sides, and also each vertex belongs to the same number of edges.

Three examples of regular polyhedra come more or less readily to mind: the cube, the tetrahedron and the octahedron. These are three of the famous five *Platonic solids;* the other two are the less obvious *icosahedron,* which has 20 triangular faces, and *dodecahedron,* which has 12 pentagonal faces. Here are pictures of the octahedron, icosahedron and dodecahedron:

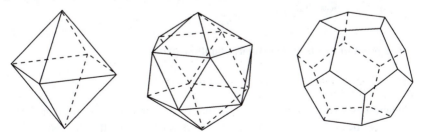

Every regular polyhedron carries five associated numbers: three are V, E, F, and the other two are n, the number of sides on a face, and r, the number of edges each vertex belongs to. We record these numbers for the Platonic solids:

	V	E	F	n	r
tetrahedron	4	6	4	3	3
cube	8	12	6	4	3
octahedron	6	12	8	3	4
icosahedron	12	30	20	3	5
dodecahedron	20	30	12	5	3

As you might have guessed from the name, the Platonic solids were known to the Greeks. They are the most symmetrical, elegant and robust of solids, so it is natural to look for more regular polyhedra. Remarkably, though perhaps disappointingly, there are no others. This fact is another theorem of the great Euler. The proof is a wonderful application of Euler's formula 9.1. Here it is.

THEOREM 9.3
The only regular convex polyhedra are the five Platonic solids.

PROOF   Suppose we have a regular polyhedron with parameters V, E, F, n and r as defined above.

First we need to show some relationships between these parameters. We shall prove first that

$$2E = nF . \tag{9.1}$$

To prove this, let us calculate the number of pairs

$$e, f$$

where e is an edge, f is a face, and e lies on f. Well, there are E possibilities for the edge e, and each lies in 2 faces f; so the number of such pairs e, f is equal to $2E$. On the other hand, there are F possibilities for the face f, and each has n edges e; so the number of such pairs e, f is also equal to nF. Therefore, $2E = nF$, proving (9.1).

Next we show that

$$2E = rV . \tag{9.2}$$

The proof of this is quite similar: count the pairs

$$v, e$$

where v is a vertex, e an edge, and v lies on e. There are E edges e, and each has 2 vertices v, so the number of such pairs v, e is $2E$; on the other hand, there are V vertices v, and each lies on r edges, so the number of such pairs is also rV. This proves (9.2).

At this point we appeal to Euler's formula 9.1:

$$V - E + F = 2 .$$

Substituting $V = \frac{2E}{r}$, $F = \frac{2E}{n}$ from (9.1) and (9.2), we obtain $\frac{2E}{r} - E + \frac{2E}{n} = 2$, hence

$$\frac{1}{r} + \frac{1}{n} = \frac{1}{2} + \frac{1}{E} . \qquad (9.3)$$

Now we know that $n \geq 3$, as a polygon must have at least 3 sides; likewise $r \geq 3$, since it is geometrically clear that in a polyhedron a vertex must belong to at least 3 edges. By (9.3), it certainly cannot be the case that both $n \geq 4$ and $r \geq 4$, since this would make the left-hand side of (9.3) at most $\frac{1}{2}$, whereas the right-hand side is more than $\frac{1}{2}$. It follows that either $n = 3$ or $r = 3$.

If $n = 3$, then (9.3) becomes

$$\frac{1}{r} = \frac{1}{6} + \frac{1}{E} .$$

The right-hand side is greater than $\frac{1}{6}$, and hence $r < 6$. Therefore, $r = 3, 4$ or 5 and $E = 6, 12$ or 30, respectively. The possible values of V, F are given by (9.1) and (9.2).

Likewise, if $r = 3$, (9.3) becomes $\frac{1}{n} = \frac{1}{6} + \frac{1}{E}$, and we argue similarly that $n = 3, 4$ or 5 and $E = 6, 12$ or 30, respectively.

We have now shown that the numbers V, E, F, n, r for a regular polyhedron must be one of the possibilities in the following table:

V	E	F	n	r
4	6	4	3	3
8	12	6	4	3
6	12	8	3	4
12	30	20	3	5
20	30	12	5	3

These are the parameter sets of the tetrahedron, cube, octahedron, icosahedron and dodecahedron, respectively. To complete the proof we now only have to show that each Platonic solid is the only regular solid with its particular parameter set. This is a simple geometric argument, and we present it just for the last parameter set — the proofs for the other sets are entirely similar.

So suppose we have a regular polyhedron with 20 pentagonal faces, each vertex lying on 3 edges. Focus on a particular vertex. At this vertex there is only one way of fitting three pentagonal faces together:

At each of the other vertices of these three pentagons, there is likewise only one way of fitting two further pentagons together. Carrying on this argument with all new vertices, we see that there is at most one way to make a regular solid with these parameters. Since the dodecahedron is such a solid, it is the only one. This completes the proof.   ∎

Exercises for Chapter 9

1. Consider a convex polyhedron, all of whose faces are squares or regular pentagons. Say there are m squares and n pentagons. Assume that each vertex lies on exactly 3 edges.

   (a) Show that for this polyhedron, the following equations hold:

   $$3V = 2E, \quad 4m + 5n = 2E, \quad m + n = F.$$

   (b) Using Euler's formula, deduce that $2m + n = 12$.

   (c) Find examples of such polyhedra for as many different values of m as you can.

2. Prove that for a convex polyhedron with V vertices, E edges and F faces, the following inequalities are true:

   $$2E \geq 3F, \quad \text{and} \quad 2E \geq 3V.$$

   Deduce using Euler's formula that

   $$2V \geq F + 4, \quad 3V \geq E + 6, \quad 2F \geq V + 4 \quad \text{and} \quad 3F \geq E + 6.$$

   Give an example of a convex polyhedron for which all these inequalities are equalities (i.e., $2V = F + 4$, etc.).

3. Draw all the connected plane graphs with 4 edges, and all the connected plane graphs with 4 vertices.

4. Critic Ivor Smallbrain has been thrown into prison for libelling the great film director Michael Loser. During one of his needlework classes in prison, Ivor is given a pile of pieces of leather in the shapes of regular pentagons and regular hexagons, and is told to sew some of these together into a convex polyhedron (which will then be used as a football). He is told that each vertex must lie on exactly 3 edges. Ivor immediately exclaims, "Then I need exactly 12 pentagonal pieces!"

   Prove that Ivor is correct.

Chapter 10

Introduction to Analysis

We saw in Chapter 2 how to prove the existence of the real number $\sqrt{2}$, and more generally of $\sqrt{n}$ for any positive integer n, by means of a clever geometrical construction. However, proving the existence of cube roots, and more generally, a real n^{th} root $x^{\frac{1}{n}}$, for any positive real number x is much harder, and relies on a much deeper study of the reals. We shall undertake such a study in this chapter. Our main focus will be the proof of the existence of nth roots. (We have already stated this result in anticipation as Proposition 5.1, and used it in later chapters.) However, I should point out that the material introduced in this chapter is a fundamental starting point for a huge area of mathematics called Analysis, which is basically the study of functions of real and complex numbers.

Upper and Lower Bounds

Our analysis will be based on the theory of *bounds* for sets of real numbers. Here is the definition.

DEFINITION   *Let S be a non-empty subset of* $\mathbb{R}$. *(So S is a set consisting of some real numbers, and* $S \neq \emptyset$*). We say that a real number u is an* upper bound *for S if*

$$s \leq u \ \text{ for all } s \in S \,.$$

Likewise, l is a lower bound *for S if*

$$s \geq l \ \text{ for all } s \in S \,.$$

Examples

(1) Let

$$S = \left\{ \frac{1}{n} \,\middle|\, n \in \mathbb{N} \right\} = \left\{ 1, \frac{1}{2}, \frac{1}{3}, \frac{1}{4}, \dots \right\} .$$

Then 1 is an upper bound for S; so are 2, 17, and indeed any number which is at least 1. Also 0 is a lower bound for S, and so is any number less than or equal to 0.

(2) If $S = \mathbb{Z}$, then S has no upper or lower bound.

(3) If

$$S = \left\{ x \,\middle|\, x \in \mathbb{Q}, \ x^2 < 2 \right\} ,$$

i.e., the set of rationals with square less than 2, then $\sqrt{2}$ is an upper bound for S, and $-\sqrt{2}$ is a lower bound.

As we see from these examples, a set can have many upper bounds. It turns out to be a fundamental question to ask whether, among all the upper bounds, there is always a least one. Let us first formally define such a thing.

DEFINITION   *Let S be a non-empty subset of $\mathbb{R}$, and suppose S has an upper bound. We say that a real number c is a* least upper bound *(abbreviated LUB), if the following two conditions hold:*
   (i) c is an upper bound for S, and
   (ii) if u is any other upper bound for S, then $u \geq c$.
Similarly, d is a greatest lower bound *(GLB) for S if*
   (a) d is a lower bound for S, and
   (b) if l is any other lower bound for S, then $l \leq d$.

Example

Let $S = \{ \frac{1}{n} \mid n \in \mathbb{N} \}$ as in Example (1) above.

We claim that 1 is a LUB for S. To see this, observe that 1 is an upper bound; and any other upper bound is at least 1, since $1 \in S$.

We also claim that 0 is a GLB for S. This is not quite so obvious. First, 0 is a lower bound. Let l be another lower bound for S. If $l > 0$, then we can find $n \in \mathbb{N}$ such that $\frac{1}{n} < l$; but $\frac{1}{n} \in S$, so this is not possible as l is a lower bound for S. Hence $l \leq 0$, which proves that 0 is a GLB for S.

The next result is absolutely fundamental to the study of the real numbers.

THEOREM 10.1

Let S be a non-empty subset of $\mathbb{R}$.

(I) *If S has a lower bound, then it has a greatest lower bound.*

(II) *If S has an upper bound, then it has a least upper bound.*

PROOF   (I) Suppose S has a lower bound. Now every member s of S has a decimal expression

$$s = s_0 \cdot s_1 s_2 s_3 \ldots$$

Since S has a lower bound, the integers s_0 occurring cannot decrease indefinitely; so there must be a smallest integer s_0 occurring; call it a_0.

Among all the members of S with decimal expressions starting with this integer a_0 (i.e., starting $a_0 \cdot s_1 s_2 \ldots$), choose one with the digit s_1 in the first decimal place as small as possible; call this digit a_1.

Similarly, among all the members of S starting $a_0 \cdot a_1 s_2 s_3 \ldots$, choose one with the digit in the second decimal place as small as possible; call this digit a_2. Carry on choosing digits $a_3, a_4, \ldots$ in this way. Now define the real number

$$d = a_0 \cdot a_1 a_2 a_3 \ldots$$

We shall show that d is a greatest lower bound for S.

First we must show that d is a lower bound. Let $s \in S$, with $s = s_0 \cdot s_1 s_2 s_3 \ldots$. If $s \neq d$, let the first decimal place where s and d disagree be the kth place. Then $s = a_0 \cdot a_1 \ldots a_{k-1} s_k \ldots$ with $s_k \neq a_k$ (possibly $k = 0$ of course). By our choice of a_k (as the least k^{th} digit occurring among members of S which start $a_0 \cdot a_1 \ldots a_{k-1} \ldots$), we must have $s_k > a_k$. This means that $s \geq d$. Hence, d is a lower bound for S.

Now we show d is a GLB. Let l be a lower bound for S, and let $l = l_0 \cdot l_1 l_2 \ldots$. We need to prove that $d \geq l$. If $d \neq l$, let the first decimal place where d and l disagree be the rth place. Then $l = a_0 \cdot a_1 \ldots a_{r-1} l_r \ldots$ with $l_r \neq a_r$. From our choice of the digits $a_0, a_1, \ldots$, we know that there is a member s of S which starts $s = a_0 \cdot a_1 \ldots a_{r-1} a_r \ldots$. Since l is a lower bound for S, we have $s \geq l$ and hence $a_r > l_r$. Therefore $d \geq l$, proving that d is a GLB for S.

This completes the proof of (I).

(II) We shall use a simple trick to deduce (II) from (I). Suppose S has an upper bound, say u. Define the set $-S$ to consist of the negatives of all members of S; that is,

$$-S = \{-s \mid s \in S\} \ .$$

As $s \leq u$ for all $s \in S$, we have $-s \geq -u$ for all $-s \in -S$, and so $-u$ is a lower bound for $-S$. Therefore by (I), the set $-S$ has a GLB; call it c.

It is now very easy to see that $-c$ is a LUB for S: first, $-s \geq c$ for all $-s \in -S$, hence $s \leq -c$ for all $s \in S$, and so $-c$ is an upper bound for S. If u is any other upper bound, then $-u$ is a lower bound for $-S$, so $-u \leq c$, which implies $u \geq -c$. Hence, $-c$ is a LUB for S, as required.   ∎

Note If a set S has a LUB, it is easy to see that it has only one LUB. We leave this to the reader (it is Exercise 2 at the end of the chapter in case you forget). So it makes sense to talk about *the* least upper bound of S. We sometimes denote this by $\text{LUB}(S)$. Likewise, a set S with a lower bound has only one GLB, denoted by $\text{GLB}(S)$.

As we have said, Theorem 10.1 underlies the whole of the theory of the real numbers, and you will see it used many times in your future study of mathematics. For now, we give one application by proving the existence of nth roots.

Existence of nth Roots

We aim now to prove the following result, already stated as Proposition 5.1.

PROPOSITION 10.1
Let n be a positive integer. If x is a positive real number, then there is exactly one positive real number y such that $y^n = x$.

The idea behind our proof of this result is very simple, but the proof itself involves quite a bit of notation, which may slightly obscure the idea at first sight. So to make everything crystal clear, we first present an example to illustrate the idea.

Example 10.1
In this example, we prove the existence of a real number c such that $c^3 = 2$. That is, we prove the existence of $2^{\frac{1}{3}}$, the real cube root of 2.

The key idea is to define the following set of real numbers:

$$S = \left\{ x \mid x \in \mathbb{R}, \ x^3 < 2 \right\}.$$

Thus, S is the set of all real numbers whose cube is less than 2.

First note that S has an upper bound; for example, 2 is an upper bound, since if $x^3 < 2$ then $x < 2$. Therefore, by Theorem 10.1, S has a least upper bound. Say

$$c = \text{LUB}(S).$$

We shall show that $c^3 = 2$. We do this by contradiction.

Assume then that $c^3 \neq 2$. Then either $c^3 < 2$ or $c^3 > 2$. We consider these two possibilities separately, in each case obtaining a contradiction.

Case 1 Assume that $c^3 < 2$. Our strategy in this case is to find a small number $\alpha > 0$ such that $(c + \alpha)^3 < 2$ still; this will mean that $c + \alpha \in S$, whereas c is an upper bound for S, a contradiction.

To find α, we argue as follows. (We have chosen to present the steps "in reverse" in order to make it clear how the argument was found.) We have

$$(c + \alpha)^3 < 2 \Leftarrow c^3 + 3c^2\alpha + 3c\alpha^2 + \alpha^3 < 2$$
$$\Leftarrow 2 - c^3 > 3c^2\alpha + 3c\alpha^2 + \alpha^3$$
$$\Leftarrow 2 - c^3 > \alpha \left(3c^2 + 3c + 1\right) \quad \text{and} \quad 0 < \alpha < 1$$

(The last inequality follows since when $0 < \alpha < 1$, we have $\alpha^2 < \alpha$ and $\alpha^3 < \alpha$.) Since $2 - c^3 > 0$ we can choose α such that

$$0 < \alpha < 1 \quad \text{and} \quad \alpha < \frac{2 - c^3}{3c^2 + 3c + 1}.$$

Then by the above implications it follows that $(c + \alpha)^3 < 2$. This leads to a contradiction, as explained before.

Case 2 Now assume that $c^3 > 2$. In this case our strategy is to find a small number $\beta > 0$ such that $(c - \beta)^3 > 2$ still. If we do this, then for $x \in S$ we have $x^3 < 2 < (c - \beta)^3$, hence $x < c - \beta$, and so $c - \beta$ is an upper bound for S. However c is the LUB of S, so this is a contradiction.

To find β, note that

$$(c - \beta)^3 > 2 \Leftarrow c^3 - 2 > 3c^2\beta - 3c\beta^2 + \beta^3$$
$$\Leftarrow c^3 - 2 > \beta(3c^2 + 3c + 1) \quad \text{and} \quad 0 < \beta < 1.$$

Since $c^3 - 2 > 0$ we can choose β such that

$$0 < \beta < 1 \quad \text{and} \quad \beta < \frac{c^3 - 2}{3c^2 + 3c + 1}.$$

Then by the above implications it follows that $(c - \beta)^3 > 2$. This leads to a contradiction, as explained before.

Thus we have reached a contradiction in both Cases 1 and 2, from which we conclude that $c^3 = 2$. In other words, c is the real cube root of 2.

Armed with the method of this example, we now proceed to the proof of Proposition 10.1. For this proof we shall require a couple of preliminary results.

LEMMA 10.1
(a) If $0 < q < \frac{1}{2}$ then $(1 - q)^n \geq 1 - nq$ for all $n \geq 1$.
(b) If $0 < q < \frac{1}{2}$ then $(1 + q)^n \leq 1 + 2^n q$ for all $n \geq 1$.

PROOF   Part (a) follows immediately from Example 8.2 in Chapter 8. And part (b) is Exercise 6 at the end of Chapter 8 (which you have, of course, done perfectly).   ∎

By the way, this result is called a "lemma" rather than a proposition, as it is just a helpful result to be used in the proof of another, more important result, and does not have very much interest of its own; the word "lemma" is used for results of this nature. The same term applies to the next result.

LEMMA 10.2
Let $y > 0$ and $0 < \alpha < \frac{1}{2}y$. Then for all $n \geq 1$, the following are true:
(a) $(y - \alpha)^n \geq y^n - ny^{n-1}\alpha$
(b) $(y + \alpha)^n \leq y^n + 2^n y^{n-1}\alpha.$

PROOF   For (a), observe that

$$(y - \alpha)^n = y^n \left(1 - \frac{\alpha}{y}\right)^n .$$

As $0 < \frac{\alpha}{y} < \frac{1}{2}$, by Lemma 10.1 the right-hand side above is greater than or equal to

$$y^n \left(1 - \frac{n\alpha}{y}\right) = y^n - ny^{n-1}\alpha ,$$

giving part (a).

To prove (b), note that by Lemma 10.1 we have $(1 + \frac{\alpha}{y})^n \leq 1 + 2^n \frac{\alpha}{y}$, whence

$$(y + \alpha)^n = y^n \left(1 + \frac{\alpha}{y}\right)^n \leq y^n \left(1 + 2^n \frac{\alpha}{y}\right) = y^n + 2^n y^{n-1}\alpha . \quad ∎$$

Proof of Proposition 10.1   Let x be a positive real number, and n a positive integer. We wish to find a real number y such that $y^n = x$. Motivated by the previous example, let us define S to be the following set of real numbers:

$$S = \left\{s \mid s \in \mathbb{R},\ s^n < x\right\} .$$

Certainly S has an upper bound: for if $x \geq 1$ then $x^n \geq x$, hence $s^n < x \leq x^n$ for all $s \in S$, so $s < x$ for all $s \in S$ and x is an upper bound for S; and if $x < 1$ then $s^n < x < 1$ for all $s \in S$, and so 1 is an upper bound. Therefore, by Theorem 10.1, S has a least upper bound; let

$$y = \text{LUB}(S) .$$

We shall show that $y^n = x$, so that y is the real number we are seeking. As in the example above, we shall do this by contradiction. So suppose that

$y^n \neq x$. Then either $y^n < x$ or $y^n > x$. We divide the argument into two cases accordingly.

Case 1 Assume first that $y^n < x$. Our strategy in this case is as in the example — to find a small $\alpha > 0$ such that $(y + \alpha)^n < x$ still; if we do this, we have $y + \alpha \in S$, whereas y is an upper bound for S, which is a contradiction.

To find α, observe that by Lemma 10.2(b),

$$(y + \alpha)^n < x \Leftarrow y^n + 2^n y^{n-1}\alpha < x \text{ and } 0 < \alpha < \frac{y}{2}$$

$$\Leftarrow x - y^n > 2^n y^{n-1}\alpha \text{ and } 0 < \alpha < \frac{y}{2}.$$

Since $x - y^n > 0$ in this case, we can choose α such that $0 < \alpha < \frac{y}{2}$ and $\alpha < \frac{x-y^n}{2^n y^{n-1}}$. Then the above implications show that $(y + \alpha)^n < x$, giving a contradiction as explained above.

Case 2 Now assume that $y^n > x$. In this case our strategy is to find a small number $\beta > 0$ such that $(y - \beta)^n > x$ still. If we do this, then for $s \in S$ we have $s^n < x < (y - \beta)^n$, hence $s < y - \beta$, and so $y - \beta$ is an upper bound for S. However, y is the LUB of S, so this is a contradiction.

To find β, note that by Lemma 10.2(a),

$$(y - \beta)^n > x \Leftarrow y^n - ny^{n-1}\beta > x \text{ and } 0 < \beta < \frac{y}{2}$$

$$\Leftarrow y^n - x > ny^{n-1}\beta \text{ and } 0 < \beta < \frac{y}{2}.$$

Since $y^n > x$ we can choose β such that $0 < \beta < \frac{y}{2}$ and $\beta < \frac{y^n-x}{ny^{n-1}}$. Then $(y - \beta)^n > x$, giving a contradiction as explained above.

We have reached a contradiction in both Cases 1 and 2, so we conclude that $y^n = x$. Finally, $y > 0$ since S contains some positive numbers and y is an upper bound for S. And y is unique, since if $y^n = z^n = x$ with $z > 0$, then $y = z$: for otherwise, either $y > z$ or $y < z$, implying that either $y^n < z^n$ or $y^n > z^n$, neither of which is true.

This completes the proof of Proposition 10.1. ∎

Exercises for Chapter 10

1. Which of the following sets S have an upper bound, and which have a lower bound? In the cases where these exist, state what the least upper bounds and greatest lower bounds are.

   (i) $S = \{-1, 3, 7, -2\}$

   (ii) $S = \{x \mid x \in \mathbb{R}$ and $|x - 3| < |x + 7|\}$

   (iii) $S = \{x \mid x \in \mathbb{R}$ and $x^3 - 3x < 0\}$

   (iv) $S = \{x \mid x \in \mathbb{N}$ and $x^2 = a^2 + b^2$ for some $a, b \in \mathbb{N}\}$.

2. Prove that if S is a set of real numbers, then S cannot have two different least upper bounds or greatest lower bounds.

3. Find the LUB and GLB of the following sets:

   (i) $\{x \mid x = 2^{-p} + 3^{-q}$ for some $p, q \in \mathbb{N}\}$

   (ii) $\{x \in \mathbb{R} \mid 3x^2 - 4x < 1\}$

   (iii) the set of all real numbers between 0 and 1 whose decimal expression contains no nines.

4. (a) Find a set of rationals having rational LUB.

   (b) Find a set of rationals having irrational LUB.

   (c) Find a set of irrationals having rational LUB.

5. Which of the following statements are true and which are false?

   (a) Every set of real numbers has a GLB.

   (b) For any real number r, there is a set of rationals having GLB equal to r.

   (c) Let $S \subseteq \mathbb{R}, T \subseteq \mathbb{R}$, and define $ST = \{st \mid s \in S, t \in T\}$, the set of all products of elements of S with elements of T. If c is the GLB of S, and d is the GLB of T, then cd is the GLB of ST.

6. Prove that the cubic equation $x^3 - x - 1 = 0$ has a real root (i.e., prove that there exists a real number c such that $c^3 - c - 1 = 0$). (*Hint:* try to find c as the LUB of a suitable set.)

Chapter 11

The Integers

In this chapter we begin to study the most basic, and also perhaps the most fascinating, number system of all — the integers. Our first aim will be to investigate factorization properties of integers. We know already that every integer greater than 1 has a prime factorization (Proposition 8.1). This was quite easy to prove using Strong Induction. A somewhat more delicate question is whether the prime factorization of an integer is always *unique* — in other words, whether, given an integer n, one can write it as a product of primes in only one way. The answer is yes; and this is such an important result that it has acquired the grandiose title of "The Fundamental Theorem of Arithmetic." We shall prove it in the next chapter, and try there to show why it is such an important result by giving some examples of its use. In this chapter we lay the groundwork for this.

We begin with a familiar definition.

DEFINITION   *Let $a, b \in \mathbb{Z}$. We say a divides b (or a is a factor of b) if $b = ac$ for some integer c. When a divides b, we write $a|b$.*

Usually, of course, given two integers a, b at random, it is unlikely that a will divide b. But we can "divide a into b" and get a quotient and a remainder:

PROPOSITION 11.1

Let a be a positive integer. Then for any $b \in \mathbb{Z}$, there are integers q, r such that
$$b = qa + r \quad and \quad 0 \leq r < a .$$

The integer q is called the quotient, and r is the remainder. For example, if $a = 17, b = 183$ then the equation in Proposition 11.1 is $183 = 10 \cdot 17 + 13$, the quotient is 10 and the remainder 13.

Proof of Proposition 11.1   Consider the rational number $\frac{b}{a}$. There is an integer q such that

$$q \le \frac{b}{a} < q + 1$$

(this is just saying $\frac{b}{a}$ lies between two consecutive integers). Multiplying through by the positive integer a, we obtain $qa \le b < (q + 1)a$, hence $0 \le b - qa < a$.

Now put $r = b - qa$. Then $b = qa + r$ and $0 \le r < a$, as required. ∎

PROPOSITION 11.2
Let $a, b, d \in \mathbb{Z}$, and suppose that $d|a$ and $d|b$. Then $d|(ma + nb)$ for any $m, n \in \mathbb{Z}$.

PROOF   Let $a = c_1 d$ and $b = c_2 d$ with $c_1, c_2 \in \mathbb{Z}$. Then for $m, n \in \mathbb{Z}$,

$$ma + nb = mc_1 d + nc_2 d = (mc_1 + nc_2)\,d \ .$$

Hence $d|(ma + nb)$. ∎

The Euclidean Algorithm

The Euclidean algorithm is a step-by-step method for calculating the common factors of two integers. First we need a definition.

DEFINITION   *Let $a, b \in \mathbb{Z}$. A common factor of a and b is an integer which divides both a and b. The highest common factor of a and b, written hcf(a, b), is the largest positive integer which divides both a and b.*

For example, hcf$(2, 3) = 1$ and hcf$(4, 6) = 2$. But how do we go about finding the highest common factor of two large numbers, say 5817 and 1428? This is what the Euclidean algorithm does for us — in a few simple, mindless steps.

Before presenting the algorithm in all its full glory, let us do an example.

Example 11.1
Here we find hcf$(5817, 1428)$ in a few mindless steps, as advertised. Write $b = 5817, a = 1428$, and let $d = $ hcf(a, b).

Step 1 Divide a into b and get a quotient and remainder:

$$5817 = 4 \cdot 1428 + 105 \ .$$

(*Deduction*: As $d|a$ and $d|b$, d also divides $a - 4b = 105$.)

Step 2 Divide 105 into 1428:

$$1428 = 13 \cdot 105 + 63 .$$

(*Deduction*: As $d|1428$ and $d|105$, d also divides 63.)

Step 3 Divide 63 into 105:

$$105 = 1 \cdot 63 + 42 .$$

(*Deduction*: $d|42$.)

Step 4 Divide 42 into 63:

$$63 = 1 \cdot 42 + 21 .$$

(*Deduction*: $d|21$.)

Step 5 Divide 21 into 42:

$$42 = 2 \cdot 21 + 0 .$$

Step 6 STOP!

We claim that $d = \text{hcf}(5817, 1428) = 21$, the last non-zero remainder in the above steps. We have already observed that $d|21$. To prove our claim, we work upwards from the last step to the first: namely, Step 5 shows that $21|42$; hence Step 4 shows that $21|63$; hence Step 3 shows that $21|105$; hence Step 2 shows that $21|1428$; hence Step 1 shows $21|5817$. Therefore, 21 divides both a and b, so $d \geq 21$. As $d|21$, it follows that $d = 21$, as claimed.

The general version of the Euclidean algorithm is really no more complicated than this example. Here it is.

Let a, b be integers. To calculate $\text{hcf}(a, b)$, we perform (mindless) steps as in the example: first divide a into b, getting a quotient q_1 and remainder r_1; then divide r_1 into a, getting remainder $r_2 < r_1$; then divide r_2 into r_1, getting remainder $r_3 < r_2$; and carry on like this until we eventually get a remainder 0 (which we must, as the r_is are decreasing and are ≥ 0). Say the remainder 0 occurs after $n + 1$ steps. Then the equations representing the steps are:

(1) $b = q_1 a + r_1$ $\qquad$ with $0 \leq r_1 < a$
(2) $a = q_2 r_1 + r_2$ $\qquad$ with $0 \leq r_2 < r_1$
(3) $r_1 = q_3 r_2 + r_3$ $\qquad$ with $0 \leq r_3 < r_2$
$\qquad \cdot \; \cdot$ $\qquad\qquad\qquad\qquad \cdot$
$\qquad \cdot \; \cdot$ $\qquad\qquad\qquad\qquad \cdot$
$\qquad \cdot \; \cdot$ $\qquad\qquad\qquad\qquad \cdot$

$(n-1)$ $r_{n-3} = q_{n-1} r_{n-2} + r_{n-1}$ with $0 \leq r_{n-1} < r_{n-2}$
$\quad (n)$ $r_{n-2} = q_n r_{n-1} + r_n$ $\qquad$ with $0 \leq r_n < r_{n-1}$
$(n+1)$ $r_{n-1} = q_{n+1} r_n + 0$

THEOREM 11.1
In the above, the highest common factor hcf(a, b) is equal to r_n, the last non-zero remainder.

PROOF   Let $d = \text{hcf}(a, b)$. We first show that $d \mid r_n$ by arguing from equation (1) downwards. By Proposition 11.2, d divides $b - q_1 a$, and hence by (1), $d \mid r_1$. Then by (2), $d \mid r_2$; by (3), $d \mid r_3$, and so on, until by (n), $d \mid r_n$.

Now we show that $d \geq r_n$ by working upwards from equation ($n + 1$). By ($n + 1$), $r_n \mid r_{n-1}$; hence by (n), $r_n \mid r_{n-2}$; hence by ($n - 1$), $r_n \mid r_{n-3}$, and so on, until by (2), $r_n \mid a$ and then by (1), $r_n \mid b$. Thus, r_n is a common factor of a and b, and so $d \geq r_n$.

We conclude that $d = r_n$, and the proof is complete.   ∎

The next result is an important consequence of the Euclidean algorithm.

PROPOSITION 11.3
If $a, b \in \mathbb{Z}$ and $d = \text{hcf}(a, b)$, then there are integers s and t such that

$$d = sa + tb .$$

PROOF   We use Equations (1),..., (n) above. By (n),

$$d = r_n = r_{n-2} - q_n r_{n-1} .$$

Substituting for r_{n-1} using Equation ($n - 1$), we get

$$d = r_{n-2} - q_n (r_{n-3} - q_{n-1} r_{n-2}) = x r_{n-2} + y r_{n-3}$$

where $x, y \in \mathbb{Z}$. Using Equation ($n - 2$) we can substitute for r_{n-2} in this (specifically, $r_{n-2} = r_{n-4} - q_{n-2} r_{n-3}$), to get

$$d = x' r_{n-3} + y' r_{n-4}$$

where $x', y' \in \mathbb{Z}$. Carrying on like this, we eventually get $d = sa + tb$ with $s, t \in \mathbb{Z}$, as required.   ∎

Example 11.2
We know by Example 11.1 that hcf($5817, 1428$) = 21. So by Proposition 11.3 there are integers s, t such that

$$21 = 5817s + 1428t .$$

Let us find such integers s, t.

To do this, we apply the method given in the proof of Proposition 11.3, using the equations in Steps 1 through 4 of Example 11.1. By Step 4,

$$21 = 63 - 42 \ .$$

Hence by Step 3,

$$21 = 63 - (105 - 63) = -105 + 2 \cdot 63 \ .$$

Hence by Step 2,

$$21 = -105 + 2(1428 - 13 \cdot 105) = 2 \cdot 1428 - 27 \cdot 105 \ .$$

Hence by Step 1,

$$21 = 2 \cdot 1428 - 27(5817 - 4 \cdot 1428) = -27 \cdot 5817 + 110 \cdot 1428 \ .$$

Thus we have found our integers s, t: $s = -27, t = 110$ will work. (But note that there are many other values of s, t which also work, for example $s = -27 + 1428, t = 110 - 5817$.)

Here is a consequence of Proposition 11.3.

PROPOSITION 11.4
If $a, b \in \mathbb{Z}$, then any common factor of a and b also divides hcf(a, b).

PROOF Let $d = $ hcf(a, b). By Proposition 11.3, there are integers s, t such that $d = sa + tb$. If m is a common factor of a and b, then m divides $sa + tb$ by Proposition 11.2, hence m divides d. ∎

We are now in a position to prove a highly significant fact about prime numbers: namely, that if a prime number p divides a product ab of two integers, then p divides one of the factors a and b.

DEFINITION *If $a, b \in \mathbb{Z}$ and hcf$(a, b) = 1$, we say that a and b are* coprime *to each other.*

For example, 17 and 1024 are coprime to each other. Note that by Proposition 11.3, if a, b are coprime to each other, then there are integers s, t such that $1 = sa + tb$.

PROPOSITION 11.5

Let $a, b \in \mathbb{Z}$.

(a) *Suppose c is an integer such that a, c are coprime to each other, and $c|ab$. Then $c|b$.*

(b) *Suppose p is a prime number and $p|ab$. Then either $p|a$ or $p|b$.*

PROOF   (a) By Proposition 11.3, there are integers s, t such that $1 = sa + tc$. Multiplying through by b gives

$$b = sab + tcb .$$

Since $c|ab$ and $c|tcb$, the right-hand side is divisible by c. Hence $c|b$.

(b) We show that if p does not divide a, then $p|b$. Suppose then that p does not divide a. As the only positive integers dividing p are 1 and p, hcf(a, p) must be 1 or p; it is not p as p does not divide a, hence hcf(a, p) = 1. Thus a, p are coprime to each other, and $p|ab$. It follows by part (a) that $p|b$, as required.   ■

This result [Proposition 11.5(b)] about prime numbers will be crucial in our proof of the uniqueness of prime factorization in the next chapter. To apply it there, we need to generalize it slightly to the case of a prime dividing a product of many factors, as follows.

PROPOSITION 11.6

Let $a_1, a_2, \ldots, a_n \in \mathbb{Z}$, and let p be a prime number. If $p|a_1 a_2 \ldots a_n$, then $p|a_i$ for some i.

PROOF   We prove this by induction. Let $P(n)$ be the statement of the proposition.

First, $P(1)$ says "if $p|a_1$ then $p|a_1$," which is trivially true.

Now suppose $P(n)$ is true. Let $a_1, a_2, \ldots, a_{n+1} \in \mathbb{Z}$, with $p|a_1 a_2 \ldots a_{n+1}$. We need to show that $p|a_i$ for some i.

Regard $a_1 a_2 \ldots a_{n+1}$ as a product ab, where $a = a_1 a_2 \ldots a_n$ and $b = a_{n+1}$. Then $p|ab$, so by Proposition 11.5(b), either $p|a$ or $p|b$. If $p|a$, that is to say $p|a_1 a_2, \ldots, a_n$, then by $P(n)$ we have $p|a_i$ for some i; and if $p|b$ then $p|a_{n+1}$. Thus, in either case, p divides one of the factors $a_1, a_2, \ldots, a_{n+1}$.

We have established that $P(n) \Rightarrow P(n + 1)$. Hence, by induction, $P(n)$ is true for all n.   ■

Exercises for Chapter 11

1. For each of the following pairs a, b of integers, find the highest common factor $d = \text{hcf}(a, b)$, and find integers s, t such that $d = sa + tb$:

    (i) $a = 17, b = 29$

    (ii) $a = 713, b = 552$

    (iii) $a = 299, b = 345$.

    In case (ii), find another pair of integers s', t' (different from your pair s, t) such that $d = s'a + t'b$.

2. A train leaves Moscow for St. Petersburg every 7 hours, on the hour. Show that on some days it is possible to catch this train at 9 A.M.

    Whenever there is a 9 A.M. train, Ivan takes it to visit his aunt Olga. How often does Olga see her nephew?

    Discuss the corresponding problem involving the train to Vladivostock, which leaves Moscow every 14 hours.

3. Show that for all positive integers n,

    $$\text{hcf}(6n + 8, 4n + 5) = 1 .$$

4. Let m, n be coprime integers, and suppose a is an integer which is divisible by both m and n. Prove that mn divides a.

    Show that the above conclusion is false if m and n are not coprime (i.e., there exists an integer a such that $m|a$ and $n|a$, but mn does not divide a).

5. Let $a, b, c \in \mathbb{Z}$. Define the highest common factor $\text{hcf}(a, b, c)$ to be the largest positive integer which divides a, b and c. Prove that there are integers s, t, u such that

    $$\text{hcf}(a, b, c) = sa + tb + uc .$$

    Find such integers s, t, u when $a = 91, b = 903, c = 1792$.

6. Let $n \geq 2$ be an integer. Prove that n is prime if and only if for every integer a, either $\text{hcf}(a, n) = 1$ or $n|a$.

Chapter 12

Prime Factorization

We have already seen in Chapter 8 (Proposition 8.1) that every integer greater then 1 is equal to a product of prime numbers, i.e., has a prime factorization. The main result of this chapter, the Fundamental Theorem of Arithmetic, tells us that this prime factorization is unique — in other words, there is essentially only one way of writing an integer as a product of primes. (In case you think this is somehow "obvious," have a look at Exercise 3 at the end of the chapter, to find an example of a number system where prime factorization is *not* unique.)

The Fundamental Theorem of Arithmetic may not seem terribly thrilling to you at first sight. However, it is in fact one of the most important properties of the integers, and has many consequences. I will endeavor to thrill you a little by giving a few such consequences after we have proved the theorem.

The Fundamental Theorem of Arithmetic

Without further ado then, let us state and prove the theorem.

THEOREM 12.1 (Fundamental Theorem of Arithmetic)
Let n be an integer with $n \geq 2$.
  (I) Then n is equal to a product of prime numbers: we have

$$n = p_1 \ldots p_k$$

where $p_1, \ldots, p_k$ are primes and $p_1 \leq p_2 \leq \ldots \leq p_k$.
  (II) This prime factorization of n is unique: in other words, if

$$n = p_1 \ldots p_k = q_1 \ldots q_l$$

where the p_is and q_is are all prime, $p_1 \leq p_2 \leq \ldots \leq p_k$ and $q_1 \leq q_2 \leq \ldots \leq q_l$, then
$$k = l \quad and \quad p_i = q_i \quad for\ all\ i = 1, \ldots, k.$$

The point about specifying that $p_1 \leq p_2 \leq \ldots \leq p_k$ is that this condition determines the order in which we write down the primes in the factorization of n. For example, 28 can be written as a product of primes in several ways: $2 \times 7 \times 2$, $7 \times 2 \times 2$ and $2 \times 2 \times 7$. But if we specify that the prime factors have to increase or stay the same, then the only factorization is $28 = 2 \times 2 \times 7$.

Proof of Theorem 12.1   Part (I) is just Proposition 8.1.

Now for the uniqueness part (II). We prove this by contradiction. So suppose there is some integer n which has two different prime factorizations, say

$$n = p_1 \ldots p_k = q_1 \ldots q_l$$

where $p_1 \leq p_2 \leq \ldots \leq p_k$, $q_1 \leq q_2 \leq \ldots \leq q_l$, and the list of primes $p_1, \ldots, p_k$ is not the same list as $q_1, \ldots, q_l$.

Now in the equation $p_1 \ldots p_k = q_1 \ldots q_l$, cancel any primes that are common to both sides. Since we are assuming the two factorizations are different, not all the primes cancel, and we end up with an equation

$$r_1 \ldots r_a = s_1 \ldots s_b ,$$

where each $r_i \in \{p_1, \ldots, p_k\}$, each $s_i \in \{q_1, \ldots, q_l\}$, and none of the r_is is equal to any of the s_is (i.e., $r_i \neq s_j$ for all i, j).

Now we obtain a contradiction. Certainly r_1 divides $r_1 \ldots r_a$, hence r_1 divides $s_1 \ldots s_b$. By Proposition 11.6, this implies that $r_1 | s_j$ for some j. However, s_j is prime, so its only divisors are 1 and s_j, and hence $r_1 = s_j$. But we know that none of the r_is is equal to any of the s_is, so this is a contradiction. This completes the proof of (II).   ∎

Of course, in the prime factorization given above in part (I) of Theorem 12.1, some of the p_is may be equal to each other. If we collect these up, we obtain a unique prime factorization of the form

$$n = p_1^{a_1} p_2^{a_2} \ldots p_m^{a_m} ,$$

where $p_1 < p_2 < \ldots < p_m$ and the a_is are positive integers.

Some Consequences of the Fundamental Theorem

First, here is an application of the Fundamental Theorem of Arithmetic which looks rather more obvious than it really is.

PROPOSITION 12.1

Let $n = p_1^{a_1} p_2^{a_2} \ldots p_m^{a_m}$, where the $p_i s$ are prime, $p_1 < p_2 < \ldots < p_m$ and the $a_i s$ are positive integers. If m divides n, then

$$m = p_1^{b_1} p_2^{b_2} \ldots p_m^{b_m} \quad \text{with} \quad 0 \le b_i \le a_i \ \text{for all } i .$$

For example, the only divisors of $2^{100} 3^2$ are the numbers $2^a 3^b$, where $0 \le a \le 100$, $0 \le b \le 2$.

Proof of Proposition 12.1   If $m|n$, then $n = mc$ for some integer c. Let $m = q_1^{c_1} \ldots q_k^{c_k}$, $c = r_1^{d_1} \ldots r_l^{d_l}$ be the prime factorizations of m, c. Then $n = mc$ gives the equation

$$p_1^{a_1} p_2^{a_2} \ldots p_m^{a_m} = q_1^{c_1} \ldots q_k^{c_k} r_1^{d_1} \ldots r_l^{d_l} .$$

By the Fundamental Theorem 12.1, the primes, and the powers to which they occur, must be identical on both sides. Hence, each q_i is equal to some p_j, and its power c_i is at most a_j. In other words, the conclusion of the proposition holds. ∎

Here is our next application.

PROPOSITION 12.2

Let n be a positive integer. Then $\sqrt{n}$ is rational if and only if n is a perfect square (i.e., $n = m^2$ for some integer m).

PROOF   The right-to-left implication is obvious: if $n = m^2$ with $m \in \mathbb{Z}$, then $\sqrt{n} = |m| \in \mathbb{Z}$ is certainly rational.

The left-to-right implication is much less clear. Suppose $\sqrt{n}$ is rational, say

$$\sqrt{n} = \frac{r}{s}$$

where $r, s \in \mathbb{Z}$. Squaring, we get $ns^2 = r^2$. Now consider prime factorizations. Each prime in the factorization of r^2 appears to an even power (since if $r = p_1^{a_1} \ldots p_k^{a_k}$ then $r^2 = p_1^{2a_1} \ldots p_k^{2a_k}$). The same holds for the primes in the factorization of s^2. Hence, by the Fundamental Theorem, each prime factor of n must also occur to an even power — say $n = q_1^{2b_1} \ldots q_l^{2b_l}$. Then $n = m^2$, where $m = q_1^{b_1} \ldots q_l^{b_l} \in \mathbb{Z}$. ∎

A similar argument applies to the rationality of cube roots, and more generally, n^{th} roots (see Exercise 2 at the end of the chapter).

Now for our final consequence of the Fundamental Theorem 12.1. Again it looks rather innocent, but in the example following the proposition we shall give a striking application of it.

In the statement, when we say a positive integer is a square (or an nth power), we mean that it is the square of an integer (or the nth power of an integer).

PROPOSITION 12.3

Let a and b be positive integers which are coprime to each other.

(a) If ab is a square, then both a and b are also squares.

(b) More generally, if ab is an nth power (for some positive integer n), then both a and b are also n^{th} powers.

PROOF　(a) Let the prime factorizations of a, b be

$$a = p_1^{d_1} \ldots p_k^{d_k}, \quad b = q_1^{e_1} \ldots q_l^{e_l}$$

(where $p_1 < \ldots < p_k$ and $q_1 < \ldots < q_l$). If ab is a square, then $ab = c^2$ for some integer c; let c have prime factorization $c = r_1^{f_1} \ldots r_m^{f_m}$. Then $ab = c^2$ gives the equation

$$p_1^{d_1} \ldots p_k^{d_k} q_1^{e_1} \ldots q_l^{e_l} = r_1^{2f_1} \ldots r_m^{2f_m}.$$

Since a and b are coprime to each other, none of the p_is are equal to any of the q_is. Hence, the Fundamental Theorem 12.1 implies that each p_i is equal to some r_j, and the corresponding powers d_i and $2f_j$ are equal; and likewise for the q_is and their powers.

We conclude that all the powers d_i, e_i are even numbers — say $d_i = 2d_i', e_i = 2e_i'$. This means that

$$a = \left(p_1^{d_1'} \ldots p_k^{d_k'} \right)^2, \quad b = \left(q_1^{e_1'} \ldots q_l^{e_l'} \right)^2,$$

so a and b are squares.

(b) The argument for (b) is the same as for (a): an equation $ab = c^n$ gives an equality

$$p_1^{d_1} \ldots p_k^{d_k} q_1^{e_1} \ldots q_l^{e_l} = r_1^{nf_1} \ldots r_m^{nf_m}.$$

The Fundamental Theorem implies that each power d_i, e_i is a multiple of n, and hence a, b are both n^{th} powers. ∎

Example 12.1

Here is an innocent little question about the integers:

Can a non-zero even square exceed a cube by 1?

(The non-zero even squares are of course the integers 4, 16, 64, 100, 144, ... and the cubes are ..., $-8, -1, 0, 1, 8, 27, ...$)

In other words, we are asking whether the equation

$$4x^2 = y^3 + 1 \tag{12.1}$$

has any solutions with x, y both non-zero integers. This is an example of a *Diophantine equation*. In general, a Diophantine equation is an equation for which the solutions are required to be integers. Most Diophantine equations are very hard, or impossible, to solve — for instance, even the equation $x^2 = y^3 + k$ has not been completely solved for all values of k. However, I have chosen a nice example, in that Equation (12.1) can be solved fairly easily (as you will see), but the solution is not totally trivial and involves use of the consequence 12.3 of the Fundamental Theorem 12.1.

Let us then go about solving Equation (12.1) for x, $y \in \mathbb{Z}$. First we rewrite it as $y^3 = 4x^2 - 1$, and then cleverly factorize the right-hand side to get

$$y^3 = (2x + 1)(2x - 1) .$$

The factors $2x + 1, 2x - 1$ are both odd integers, and their highest common factor divides their difference, which is 2. Hence

$$\mathrm{hcf}(2x + 1, 2x - 1) = 1 .$$

Thus, $2x + 1$ and $2x - 1$ are coprime to each other, and their product is y^3, a cube. By Proposition 12.3(b), it follows that $2x + 1$ and $2x - 1$ are themselves both cubes. However, from the list of cubes ..., $-8, -1, 0, 1, 8, 27, ...$ it is apparent that the only two cubes which differ by 2 are 1, -1. Therefore, $x = 0$ and we have shown that the only even square which exceeds a cube by 1 is 0. In other words, there are no non-zero such squares.

Exercises for Chapter 12

1. Suppose that n is a positive integer such that $p = 2^n - 1$ is prime. (The first few such primes are $3, 7, 31, \ldots$.) Define

$$N = 2^{n-1} p \, .$$

List all positive integers which divide N. Prove that the sum of all these divisors, including 1 but not N itself, is equal to N.

A positive integer which is equal to the sum of all its divisors (including 1 but not itself) is called a *perfect* number. Write down four perfect numbers.

2. (a) Prove that $2^{\frac{1}{3}}$ and $3^{\frac{1}{3}}$ are irrational.

(b) Let m and n be positive integers. Prove that $m^{\frac{1}{n}}$ is rational if and only if m is an nth power (i.e., $m = c^n$ for some integer c).

3. Let E be the set of all positive even integers. We call a number e in E "prima" if e cannot be expressed as a product of two other members of E.

   (i) Show that 6 is prima but 4 is not.

   (ii) What is the general form of a prima in E?

   (iii) Prove that every element of E is equal to a product of primas.

   (iv) Give an example to show that E does not satisfy a "unique prima factorization theorem" (i.e., find an element of E which has two different factorizations as a product of primas).

4. Find all solutions $x, y \in \mathbb{Z}$ to the following Diophantine equations:

(a) $x^2 = y^3$

(b) $x^2 = y^4 - 77$

(c) $x^3 = 4y^2 + 4y - 3$.

5. Languishing in his prison cell, critic Ivor Smallbrain is dreaming. In his dream he is on the Pacific island of Nefertiti, eating coconuts on a beach by a calm blue lagoon. Suddenly the King of Nefertiti approaches him, saying, "Your head will be chopped off unless you answer this riddle: Is it possible for the sixth power of an integer to exceed the fifth power of another integer by 16?" Feverishly, Ivor writes some calculations in the sand, and eventually answers, "Oh Great King, no it is not possible." The king rejoinders, "You are correct, but you will be beheaded anyway." The

executioner's axe is just coming down when Ivor wakes up. He wonders whether his answer to the king was really correct.

Prove that Ivor was indeed correct.

Chapter 13

More on Prime Numbers

As you are probably beginning to appreciate, the prime numbers are fundamental to our understanding of the integers. In this chapter we will discuss a few basic results concerning the primes, and also hint at the vast array of questions, some solved, some unsolved, in current research into prime numbers.

The first few primes are

$$2, 3, 5, 7, 11, 13, 17, 19, 23, 29, 31, 37, 41, 43, \ldots$$

It is quite simple to carry on a long way with this list, particularly if you have a computer at hand. How would you do this? The easiest way is probably to test each successive integer n for primality, by checking, for each prime $p \leq \sqrt{n}$, whether p divides n (such primes p will of course already be in your list). If none of these primes p divides n, then n is prime (see Exercise 1 at the end of the chapter).

Probably the first and most basic question to ask is: Does this list ever stop? In other words, is there a *largest* prime number, or does the list of primes go on forever? The answer is provided by the following famous theorem of Euclid (300 BC).

THEOREM 13.1
There are infinitely many prime numbers.

PROOF   This is one of the classic proofs by contradiction. Assume the result is false, i.e., there are only finitely many primes. This means that we can make a finite list

$$p_1, p_2, p_3, \ldots, p_n$$

of *all* the prime numbers. Now define a positive integer

$$N = p_1 p_2 p_3 \ldots p_n + 1.$$

By Proposition 8.1, N is equal to a product of primes, say $N = q_1 \ldots q_r$ with all q_i prime. As q_1 is prime, it belongs to the above list of all primes, so $q_1 = p_i$ for some i.

Now q_1 divides N, hence p_i divides N. Also p_i divides $p_1 p_2 \ldots p_n$, which is equal to $N - 1$. Thus, p_i divides both N and $N - 1$. But this implies that p_i divides the difference between these numbers, namely 1. This is a contradiction.

∎

Theorem 13.1 is of course not the end of the story about the primes — it is really the beginning. A natural question to ask which flows from the theorem is: What *proportion* of all positive integers are prime? On the face of it this question makes no sense, as the integers and the primes are both infinite sets. But one can make a sensible question by asking

Given a positive integer n, how many of the numbers 1, 2, 3, ..., *n are prime?*

Is there any reason to expect to be able to answer this question? On the face of it, no. If you stare at a long list of primes, you will see that the sequence is very irregular, and it is very difficult to see any pattern at all in it. (See, for example, Exercise 3 at the end of the chapter.) Why on earth should there then be a nice formula for the number of primes up to n?

The amazing thing is that there *is* such a formula, albeit an "asymptotic" one (I will explain this word later). The great Gauss, by calculating a lot with lists of primes (and also by having a lot of brilliant thoughts) formed the incredible conjecture (i.e., informed guess) that the number of primes up to n should be pretty close to the formula

$$\frac{n}{\log_e n}.$$

To understand this a little, compare the number of primes up to, say, 10^6 — namely, 78498 — with the value of $\frac{10^6}{\log 10^6}$ — namely, 72382.4. The *difference* between these two numbers, about 6000, appears to be quite large; but their *ratio* is 1.085, quite close to 1. It was on the ratio, rather than the difference, that Gauss concentrated his mind: his conjecture was that the ratio of the number of primes up to n and the expression $\frac{n}{\log_e n}$ should get closer and closer to 1 as n gets larger and larger. (Formally, this ratio *tends to* 1 *as n tends to infinity*.)

Gauss did not actually manage to prove his conjecture. The world had to wait until 1896, when a Frenchman, Hadamard, and a Belgian, de la Vallée-Poussin, both produced proofs of what is now known as the Prime Number Theorem:

THEOREM 13.2

For a positive integer n, let $\pi(n)$ be the number of primes up to n. Then the ratio of $\pi(n)$ and $\frac{n}{\log_e n}$ tends to 1 as n tends to infinity (i.e., the ratio can be made as close as we like to 1 provided n is large enough).

The proof of this result uses some quite sophisticated tools of Analysis. Nevertheless, if you are lucky you might get the chance to see a proof in an undergraduate course later in your studies — in other words, it is not *that* difficult!

You should not think that every question about the primes can be answered (if not by you, then by some expert or other). On the contrary, many basic questions about the primes are unsolved to this day, despite being studied for many years. Let me finish this chapter by mentioning a couple of the most famous such problems.

The Goldbach conjecture If you do some calculations, or program your computer, you will find that any reasonably small even positive integer greater than 2 can be expressed as a sum of two primes. For example,

$$10 = 7 + 3, \; 50 = 43 + 7, \; 100 = 97 + 3, \; 8000 = 3943 + 4057$$

and so on. Based on this evidence, it seems reasonable to conjecture that *every* even positive integer is the sum of two primes. This is the Goldbach conjecture, and it is unsolved to this day.

The twin prime conjecture If p and $p + 2$ are both prime numbers, we call them *twin primes*. For example, here are some twin primes:

$$3, 5; \quad 5, 7; \quad 11, 13; \quad 71, 73; \quad 1997, 1999.$$

If you stare at a list of prime numbers, you will find many pairs of twin primes, getting larger and larger. One feels that there should be infinitely many twin primes, and indeed, that statement is known as the twin prime conjecture. Can one prove the twin prime conjecture using a proof like Euclid's in Theorem 13.1? Unfortunately not — indeed, no one has come up with any sort of proof, and the conjecture remains unsolved to this day.

Exercises for Chapter 13

1. Let n be an integer with $n \geq 2$. Suppose that for every prime $p \leq \sqrt{n}$, p does not divide n. Prove that n is prime.

   Is 221 prime? Is 223 prime?

2. For a positive integer n, define $\phi(n)$ to be the number of positive integers $a < n$ such that $\text{hcf}(a, n) = 1$. (For example, $\phi(2) = 1$, $\phi(3) = 2$, $\phi(4) = 2$.)

   Work out $\phi(n)$ for $n = 5, 6, \ldots, 10$.

   If p is a prime, show that $\phi(p) = p - 1$, and more generally, that $\phi(p^r) = p^r - p^{r-1}$.

3. There has been quite a bit of work over the years on trying to find a nice formula that takes many prime values. For example, $x^2 + x + 41$ is prime for all integers x such that $-40 \leq x < 40$. (You may like to check this!) However:

   Find an integer x such that $x^2 + x + 41$ is not prime.

4. Use the idea of the proof of Euclid's theorem 13.1 to prove that there are infinitely many primes of the form $4k + 1$ (where k is an integer).

5. On his release from prison, critic Ivor Smallbrain rushes out to see the latest film, *Prime and Prejudice.* During the film Ivor attempts to think of ten consecutive positive integers, none of which is prime. He fails.

   Can you help Ivor? More generally, show that for any $n \in \mathbb{N}$ there is a sequence of n consecutive positive integers, none of which is prime. (Hence, there are arbitrarily large "gaps" in the sequence of primes.)

Chapter 14

Congruence of Integers

In this chapter we introduce another method for studying the integers, called congruence. Let us go straight into the definition.

DEFINITION   *Let m be a positive integer. For $a, b \in \mathbb{Z}$, if m divides $b - a$ we write $a \equiv b \bmod m$, and say a is congruent to b modulo m.*

For example,

$$5 \equiv 1 \bmod 2, \quad 12 \equiv 17 \bmod 5, \quad 91 \equiv -17 \bmod 12, \quad 531 \not\equiv 0 \bmod 4 \,.$$

PROPOSITION 14.1
Every integer is congruent to exactly one of the numbers $0, 1, 2, \ldots, m - 1$ modulo m.

PROOF   Let $x \in \mathbb{Z}$. By Proposition 11.1, there are integers q, r such that

$$x = qm + r \quad \text{with } 0 \leq r < m \,.$$

Then $x - r = qm$, so m divides $x - r$, and hence by the above definition, $x \equiv r \bmod m$. Since r is one of the numbers $0, 1, 2, \ldots, m - 1$, the proposition follows.   ∎

Examples
(1) Every integer is congruent to 0 or 1 modulo 2. Indeed, all even integers are congruent to 0 modulo 2, and all odd integers to 1 modulo 2.

(2) Every integer is congruent to 0, 1, 2 or 3 modulo 4. More specifically, every even integer is congruent to 0 or 2 modulo 4, and every odd integer to 1 or 3 modulo 4.

(3) My clock is now showing the time as 2.00 A.M. What time will it be showing in 4803 hours? Since $4803 \equiv 3 \bmod 24$, it will be showing a time 3

hours later than the current time, i.e., 5.00 A.M. (but I hope I will not be awake to see it).

The next result will be quite useful for our later work involving manipulation of congruences.

PROPOSITION 14.2
Let m be a positive integer. The following are true, for all $a, b, c \in \mathbb{Z}$:
(1) $a \equiv a$ mod m,
(2) if $a \equiv b$ mod m then $b \equiv a$ mod m,
(3) if $a \equiv b$ mod m and $b \equiv c$ mod m, then $a \equiv c$ mod m.

PROOF   (1) Since $m|0$ we have $m|a - a$, hence $a \equiv a$ mod m.
(2) If $a \equiv b$ mod m then $m|b - a$, so $m|a - b$, hence $b \equiv a$ mod m.
(3) If $a \equiv b$ mod m and $b \equiv c$ mod m, then $m|b - a$ and $m|c - b$; say $b - a = km$, $c - b = lm$. Then $c - a = (k + l)m$, so $m|c - a$, and hence $a \equiv c$ mod m.   ∎

Arithmetic with Congruences

Congruence is a notation which conveniently records various divisibility properties of integers. This notation comes into its own when we do arithmetic with congruences, as we show is possible in the next two results. The first shows that congruences modulo m can be added and multiplied.

PROPOSITION 14.3
Suppose $a \equiv b$ mod m and $c \equiv d$ mod m. Then

$$a + c \equiv b + d \text{ mod } m \quad and \quad ac \equiv bd \text{ mod } m .$$

PROOF   We are given that $m|b-a$ and $m|d-c$. Say $b-a = km$, $d-c = lm$, where $k, l \in \mathbb{Z}$. Then

$$(b + d) - (a + c) = (k + l)m$$

and hence $a + c \equiv b + d$ mod m. And

$$bd - ac = (a + km)(c + lm) - ac = m(al + ck + klm)$$

which implies that $ac \equiv bd$ mod m.   ∎

PROPOSITION 14.4

If $a \equiv b \mod m$, and n is a positive integer, then

$$a^n \equiv b^n \mod m \ .$$

PROOF   We prove this by induction. Let $P(n)$ be the statement of the proposition. Then $P(1)$ is obviously true.

Now suppose $P(n)$ is true, so $a^n \equiv b^n \mod m$. As $a \equiv b \mod m$, we can use Proposition 14.3 to multiply these congruences and get $a^{n+1} \equiv b^{n+1} \mod m$, which is $P(n+1)$. Hence, $P(n)$ is true for all n by induction.   ∎

These results give us some powerful methods for using congruences, as we shall now attempt to demonstrate with a few examples.

Example 14.1
Find the remainder r (between 0 and 6) that we get when we divide 6^{82} by 7.

Answer We start with the congruence $6 \equiv -1 \mod 7$. By Proposition 14.4, we can raise this to the power 82, to get $6^{82} \equiv (-1)^{82} \mod 7$, hence $6^{82} \equiv 1 \mod 7$. This means that 7 divides $6^{82} - 1$, i.e., $6^{82} = 7q + 1$ for some $q \in \mathbb{Z}$. Hence, the remainder is 1.

Example 14.2
Find the remainder r (between 0 and 12) that we get when we divide 6^{82} by 13.

Answer This is not quite so easy as the previous example. We employ a general method, which involves "successive squaring" of the congruence $6 \equiv 6 \mod 13$. Squaring once, we get $6^2 \equiv 36 \mod 13$; since $36 \equiv -3 \mod 13$, Proposition 14.2(3) gives $6^2 \equiv -3 \mod 13$. Successive squaring like this yields:

$$6^2 \equiv -3 \mod 13 \ ,$$
$$6^4 \equiv 9 \mod 13 \ ,$$
$$6^8 \equiv 3 \mod 13 \ ,$$
$$6^{16} \equiv 9 \mod 13 \ ,$$
$$6^{32} \equiv 3 \mod 13 \ ,$$
$$6^{64} \equiv 9 \mod 13 \ .$$

Now $6^{82} = 6^{64} 6^{16} 6^2$. Multiplying the above congruences for 6^{64}, 6^{16} and 6^2, we get

$$6^{82} \equiv 9 \cdot 9 \cdot (-3) \mod 13 \ .$$

Now $9 \cdot (-3) = -27 \equiv -1 \bmod 13$, so $6^{82} \equiv -9 \equiv 4 \bmod 13$. Hence, the required remainder is 4.

The method given in this example is called the "method of successive squares," and always works to yield the congruence of a large power, given some effort. Often this effort can be reduced with some clever trickery, as in the previous example.

Example 14.3

Show that no integer square is congruent to 2 modulo 3. (In other words, the sequence $2, 5, 8, 11, 14, 17, \ldots$ contains no squares.)

Answer Consider an integer square n^2 (where $n \in \mathbb{Z}$). By Proposition 14.1, n is congruent to 0, 1 or 2 modulo 3. If $n \equiv 0 \bmod 3$, then by Proposition 14.4, $n^2 \equiv 0 \bmod 3$; if $n \equiv 1 \bmod 3$, then $n^2 \equiv 1 \bmod 3$; and if $n \equiv 2 \bmod 3$, then $n^2 \equiv 4 \bmod 3$, hence [using 14.2(3)] $n^2 \equiv 1 \bmod 3$. This shows that integer squares are congruent to 0 or 1 modulo 3.

Example 14.4

Show that every odd integer square is congruent to 1 modulo 4.

Answer This is similar to the previous example. Let n be an odd integer. Then n is congruent to 1 or 3 modulo 4, so n^2 is congruent to 1 or 9 modulo 4, hence to 1 modulo 4.

Example 14.5

The "rule of 3" You may have come across a simple rule for testing whether an integer is divisible by 3: add up its digits, and if the sum is divisible by 3 then the integer is divisible by 3. Here is a quick explanation of why this rule works.

Let n be an integer, with digits $a_r a_{r-1} \ldots a_0$, so

$$n = a_0 + 10a_1 + 10^2 a_2 + \cdots + 10^r a_r .$$

Now $10 \equiv 1 \bmod 3$, hence by Proposition 14.4, $10^k \equiv 1 \bmod 3$ for any positive integer k. Multiplying this by the congruence $a_k \equiv a_k \bmod 3$ gives $10^k a_k \equiv a_k \bmod 3$. It follows that

$$n \equiv a_0 + a_1 + \cdots + a_r \bmod 3 .$$

Hence, $n \equiv 0 \bmod 3$ if and only if the sum of its digits $a_0 + \cdots + a_r \equiv 0 \bmod 3$. This is the "rule of 3."

The same method proves the "rule of 9": an integer is divisible by 9 if and only if the sum of its digits is divisible by 9. There is also a "rule of 11," which is not quite so obvious: an integer n with digits $a_r \ldots a_1 a_0$ is divisible by 11 if and only if the expression $a_0 - a_1 + a_2 - \cdots + (-1)^r a_r$ is divisible by 11. Proving this is Exercise 3 at the end of the chapter.

Congruence Equations

Let m be a positive integer and let $a, b \in \mathbb{Z}$. Consider the equation

$$ax \equiv b \bmod m$$

to be solved for $x \in \mathbb{Z}$. Such an equation is called a linear congruence equation. When does such an equation have a solution?

Examples
(1) Consider the congruence equation

$$4x \equiv 2 \bmod 28 .$$

If $x \in \mathbb{Z}$ is a solution to this, then $4x = 2 + 28n$ for some integer n, which is impossible since the left-hand side is divisible by 4, whereas the right-hand side is not. So this congruence equation has no solutions.

(2) Now consider the equation

$$13x \equiv 2 \bmod 31 .$$

We shall show that this equation has a solution. Observe that $\text{hcf}(13, 31) = 1$, hence by Proposition 11.3, there are integers s, t such that

$$1 = 13s + 31t .$$

Therefore, $13s = 1 - 31t$, which means that $13s \equiv 1 \bmod 31$. Multiplying this congruence by 2, we get

$$13 \cdot (2s) \equiv 2 \bmod 31 .$$

In other words, $x = 2s$ is a solution to the original congruence equation.

Here is a general result telling us exactly when linear congruence equations have solutions.

PROPOSITION 14.5
The congruence equation
$$ax \equiv b \bmod m$$
has a solution $x \in \mathbb{Z}$ if and only if $\text{hcf}(a, m)$ *divides b.*

PROOF   Write $d = \text{hcf}(a, m)$. First let us prove the left-to-right implication. So suppose the equation has a solution $x \in \mathbb{Z}$. Then $ax = qm + b$ for some integer q. Since $d|a$ and $d|m$, it follows that $d|b$.

Now for the right-to-left implication. Suppose $d|b$, say $b = kd$. By Proposition 11.3, there are integers s, t such that $d = sa + tm$. Multiplying through by k gives $b = kd = k(sa + tm)$. Hence,

$$aks = b - ktm \equiv b \bmod m \ .$$

In other words, $x = ks$ is a solution to the congruence equation.   ∎

Exercises for Chapter 14

1. (a) Find r with $0 \le r \le 10$ such that $7^{137} \equiv r \bmod 11$.

   (b) Find the last two digits of 3^{124} (when expressed in decimal notation).

   (c) Show that there is a multiple of 21 which has 241 as its last three digits.

2. For each of the following congruence equations, either find a solution $x \in \mathbb{Z}$ or show that no solution exists:

   (a) $99x \equiv 18 \bmod 30$

   (b) $91x \equiv 84 \bmod 143$

   (c) $x^2 \equiv 2 \bmod 5$

   (d) $x^2 + x + 1 \equiv 0 \bmod 5$

   (e) $x^2 + x + 1 \equiv 0 \bmod 7$.

3. (a) Prove the "rule of 9": an integer is divisible by 9 if and only if the sum of its digits is divisible by 9.

   (b) Prove the "rule of 11" stated in Example 14.5. Use this rule to decide in your head whether the number 82918073579 is divisible by 11.

4. Let p be a prime number, and let a be an integer which is not divisible by p. Prove that the congruence equation $ax \equiv 1 \bmod p$ has a solution $x \in \mathbb{Z}$.

5. Show that every square is congruent to 0,1 or -1 modulo 5, and is congruent to 0,1 or 4 modulo 8.

   Suppose n is a positive integer such that both $2n + 1$ and $3n + 1$ are squares. Prove that n is divisible by 40.

   Find a value of n such that $2n + 1$ and $3n + 1$ are squares. Can you find another value? (Calculators allowed!)

6. It is Friday, June 11, 1999. Ivor Smallbrain is watching the famous movie *From Here to Infinity*. He is bored, and idly wonders what day of the week it will be on the same date in 1000 years' time (i.e., on June 11th, 2999). He decides it will again be a Friday.

   Is Ivor right? And what has this question got to do with congruence?

Chapter 15

Counting and Choosing

Mathematics has many tools for counting. We shall give some simple methods in this chapter. These lead us to binomial coefficients and then to the Binomial theorem and the Multinomial theorem.

Let us begin with an example.

Examples

A security system uses passwords consisting of two letters followed by two digits. How many different passwords are possible?

Answer The number of choices for each letter is 26 and for each digit is 10. We claim the answer is the product of these numbers, namely

$$26 \times 26 \times 10 \times 10 \,.$$

Here is the justification for this claim. Let N be the number of possible passwords, and let a typical password be $\alpha\beta\gamma\delta$, where α, β are letters and γ, δ are digits. For each choice of $\alpha\beta\gamma$, there are 10 passwords $\alpha\beta\gamma\delta$ (one for each of the 10 possibilities for δ). Thus,

$$N = 10 \times \text{ number of choices for } \alpha\beta\gamma \,.$$

Likewise, for each choice of $\alpha\beta$, there are 10 possibilities for $\alpha\beta\gamma$, so

$$N = 10 \times 10 \times \text{ number of choices for } \alpha\beta \,.$$

For each choice of α there are 26 possibilities for $\alpha\beta$, so

$$N = 10 \times 10 \times 26 \times \text{ number of choices for } \alpha = 10 \times 10 \times 26 \times 26 \,.$$

The argument used in the above example shows the following:

THEOREM 15.1 *Multiplication Principle*

Let P be a process which consists of n stages, and suppose that for each r, the rth stage can be carried out in a_r ways. Then P can be carried out in $a_1 a_2 \ldots a_n$ ways.

In the above example, there are four stages, with $a_1 = a_2 = 26$, $a_3 = a_4 = 10$.

Here is another example which is not quite so simple.

Examples

Using the digits $1, 2, \ldots, 9$, how many even two-digit numbers are there with two different digits?

Answer Let us consider the following two-stage process to pick such an even integer. The first stage is to choose the *second* digit (2, 4, 6 or 8); it can be done in four ways. The second stage is to choose the first digit, which can be done in eight ways. Hence, by the Multiplication Principle, the answer to the question is 32.

Notice that we would have had trouble if we had carried out the process the other way around and first chosen the *first* digit, because then the number of ways of choosing the second digit would depend on whether the first digit was even or odd.

Here is another application of the Multiplication Principle.

PROPOSITION 15.1

Let S be a set consisting of n elements. Then the number of different arrangements of the elements of S in order is n! (recall $n! = n(n-1)(n-2)\ldots 2 \cdot 1$).

For example, if $S = \{a, b, c\}$ then the different arrangements of the elements in order are

$$abc, \ acb, \ bac, \ bca, \ cab, \ cba \ .$$

As predicted by the proposition, there are $6 = 3!$ of them.

PROOF   Choosing an arrangement is an n-stage process. First choose the first element (n possibilities); then choose the second element — this can be any of the remaining $n-1$ elements, so there are $n-1$ possibilities; then the third, for which there are $n-2$ possibilities; and so on. Hence by the Multiplication Principle, the total number of arrangements is $n(n-1)(n-2)\ldots 2 \cdot 1 = n!$.

∎

Binomial Coefficients

We now introduce some numbers which are especially useful in counting arguments.

DEFINITION   Let n be a positive integer, and r an integer such that $0 \leq r \leq n$. Define

$$\binom{n}{r}$$

(called "n choose r") to be the number of r-element subsets of $\{1, 2, \ldots, n\}$.

For example, the 2-element subsets of $\{1, 2, 3, 4\}$ are $\{1, 2\}$, $\{1, 3\}$, $\{1, 4\}$, $\{2, 3\}$, $\{2, 4\}$, $\{3, 4\}$, and so

$$\binom{4}{2} = 6 .$$

PROPOSITION 15.2
We have

$$\binom{n}{r} = \frac{n!}{r!(n-r)!} .$$

PROOF   Let $S = \{1, 2, \ldots, n\}$ and count the arrangements of S in order as follows:

Stage 1 : Choose an r-element subset T of S: there are $\binom{n}{r}$ choices.
Stage 2 : Choose an arrangement of T: by 15.1 there are $r!$ choices.
Stage 3 : Choose an arrangement of the remaining $n - r$ elements of S: there are $(n - r)!$ choices.

By the Multiplication Principle, the total number of arrangements of S is equal to the product of these three numbers. Hence by Proposition 15.1,

$$n! = \binom{n}{r} \times r! \times (n-r)!$$

and the result follows from this.   ∎

Another way of expressing the conclusion of Proposition 15.2 is

$$\binom{n}{r} = \frac{n(n-1)\ldots(n-r+1)}{r(r-1)\ldots 2 \cdot 1} .$$

Some useful particular cases of this are:

$$\binom{n}{0} = \binom{n}{n} = 1$$

(there is only one 0-element and one n-element subset of S!), and

$$\binom{n}{1} = n, \quad \binom{n}{2} = \frac{n(n-1)}{2}, \quad \binom{n}{3} = \frac{n(n-1)(n-2)}{6}.$$

Examples

Liebeck has taught the same course for the last 16 years, and tells 3 jokes each year. He never tells the same set of 3 jokes twice. At least how many jokes does Liebeck know? When will he have to tell a new one?

Answer Suppose Liebeck knows n jokes. Then $\binom{n}{3}$ must be at least 16. Since $\binom{5}{3} = 10$ and $\binom{6}{3} = 20$, it follows that $n \geq 6$. So Liebeck knows at least 6 jokes, and will have to tell a seventh in 5 years' time (i.e., in the 21st year of giving the course — assuming he has not dropped dead by then).

The numbers $\binom{n}{r}$ are known as *binomial coefficients*. This is because of the following famous theorem.

THEOREM 15.2 Binomial Theorem

Let n be a positive integer, and let a, b be real numbers. Then

$$(a+b)^n = \sum_{r=0}^{n} \binom{n}{r} a^{n-r} b^r$$

$$= a^n + na^{n-1}b + \binom{n}{2} a^{n-2}b^2 + \cdots$$

$$+ \binom{n}{r} a^{n-r} b^r + \cdots + \binom{n}{n-1} ab^{n-1} + b^n.$$

PROOF   Consider

$$(a+b)^n = (a+b)(a+b)\ldots(a+b).$$

When we multiply this out to get a term $a^{n-r}b^r$, we choose the b from r of the brackets and the a from the other $n - r$ brackets. So the number of ways of getting $a^{n-r}b^r$ is the number of ways of choosing r brackets from n, hence is $\binom{n}{r}$. In other words, the coefficient of $a^{n-r}b^r$ is $\binom{n}{r}$. The theorem follows.

Here are the first few cases of the Binomial Theorem:

$$(a + b)^0 = 1$$
$$(a + b)^1 = a + b$$
$$(a + b)^2 = a^2 + 2ab + b^2$$
$$(a + b)^3 = a^3 + 3a^2b + 3ab^2 + b^3$$
$$(a + b)^4 = a^4 + 4a^3b + 6a^2b^2 + 4ab^3 + b^4$$
$$(a + b)^5 = a^5 + 5a^4b + 10a^3b^2 + 10a^2b^3 + 5ab^4 + b^5 .$$

There are one or two patterns to observe about this. First, each expression is symmetrical about the center; this is due to the fact, obvious from Proposition 15.2, that

$$\binom{n}{r} = \binom{n}{n - r}$$

(for example $\binom{5}{2} = \binom{5}{3} = 10$). Rather less obvious is the fact that if we write down the coefficients in the above expressions in the following array, known as *Pascal's triangle:*

$$
\begin{array}{c}
1 \\
1\ 1 \\
1\ 2\ 1 \\
1\ 3\ 3\ \ 1 \\
1\ 4\ 6\ \ 4\ \ 1 \\
1\ 5\ 10\ 10\ 5\ 1
\end{array}
$$

then you can see that each entry is the sum of the one above it and the one to the left of that. This is explained by the equality

$$\binom{n + 1}{r} = \binom{n}{r} + \binom{n}{r - 1}$$

which you are asked to prove in Exercise 2 at the end of the chapter.

Putting $a = 1$ and $b = x$ in the Binomial Theorem, we obtain the following consequence.

PROPOSITION 15.3
For any positive integer n,

$$(1 + x)^n = \sum_{r=0}^{n} \binom{n}{r} x^r .$$

Putting $x = \pm 1$ in this, we get the interesting equalities

$$\sum_{r=0}^{n} \binom{n}{r} = 2^n , \qquad \sum_{r=0}^{n} (-1)^r \binom{n}{r} = 0 .$$

Multinomial Coefficients

Suppose eight students (call them $1, 2, 3, \ldots, 8$) are to be assigned to three projects (call them A, B, C); project A requires 4 students, project B requires 2, and project C also requires 2. In how many ways can the students be assigned to the projects?

To answer this, we could list all possible assignments like this:

A	B	C
1, 2, 3, 4	5, 6	7, 8
1, 2, 3, 4	7, 8	5, 6
1, 2, 3, 4	5, 7	6, 8
1, 2, 3, 4	6, 8	5, 7
1, 2, 3, 5	4, 6	7, 8
.	.	.
.	.	.

However, the deadline for the projects will probably have passed by the time we have finished writing down the complete list (in fact there are 420 possible assignments). We need a nice way of counting such things.

Each assignment is what is called an *ordered partition* of the set $\{1, 2, \ldots, 8\}$ into subsets A, B, C of sizes 4, 2, 2. Here is the general definition of such a thing.

DEFINITION   *Let n be a positive integer, and let* $S = \{1, 2, \ldots, n\}$. *A partition of S is a collection of subsets* $S_1, \ldots, S_k$ *such that each element of S lies in exactly one of these subsets. The partition is* ordered *if we take account of the order in which the subsets are written.*

The point about the order is that, for instance in the above example, the ordered partition

$$\{1, 2, 3, 4\} \quad \{5, 6\} \quad \{7, 8\}$$

is different from the ordered partition

$$\{1, 2, 3, 4\} \quad \{7, 8\} \quad \{5, 6\}$$

even though the subsets involved are the same in both cases.

If $r_1, \ldots, r_k$ are non-negative integers such that $n = r_1 + \cdots + r_k$, we denote the total number of ordered partitions of $S = \{1, 2, \ldots, n\}$ into subsets $S_1, \ldots, S_k$ of sizes $r_1, \ldots, r_k$ by the symbol

$$\binom{n}{r_1, \ldots, r_k}.$$

Examples

(1) The number of possible project assignments in the example above is

$$\binom{8}{4, 2, 2}.$$

(2) If Alfred, Barney, Cedric and Dugald play bridge, the total number of different possible hands which can be dealt is

$$\binom{52}{13, 13, 13, 13}.$$

(3) It is rather clear that

$$\binom{n}{r, n - r} = \binom{n}{r}.$$

PROPOSITION 15.4

We have

$$\binom{n}{r_1, \ldots, r_k} = \frac{n!}{r_1! r_2! \ldots r_k!}.$$

PROOF   We count the $n!$ arrangements of $S = \{1, 2, \ldots, n\}$ in stages as follows:

Stage 0: Choose an ordered partition of S into subsets $S_1, \ldots, S_k$ of sizes $r_1, \ldots, r_k$; the number of ways of doing this is

$$\binom{n}{r_1, \ldots, r_k}.$$

Stage 1 : Choose an arrangement of S_1: there are $r_1!$ choices.

Stage 2 : Choose an arrangement of S_2: there are $r_2!$ choices.

And so on, until

Stage k : Choose an arrangement of S_k: there are $r_k!$ choices.

By the Multiplication Principle, we conclude that

$$n! = \binom{n}{r_1, \ldots, r_k} r_1! \ldots r_k!$$

The result follows.   ∎

Examples

(1) The number of project assignments in the first example is

$$\binom{8}{4,\,2,\,2} = \frac{8!}{4!2!2!} = 420\,.$$

(2) The total number of bridge hands, namely

$$\binom{52}{13,\,13,\,13,\,13},$$

is approximately 5.365×10^{28}, quite a large number.

The numbers $\binom{n}{r_1,\ldots,r_k}$ are called *multinomial coefficients,* for the following reason.

THEOREM 15.3 Multinomial Theorem

Let n be a positive integer, and let $x_1,\ldots,x_k$ be real numbers. Then the expansion of $(x_1 + \cdots + x_k)^n$ is the sum of all terms of the form

$$\binom{n}{r_1,\,\ldots,\,r_k} x_1^{r_1} \ldots x_k^{r_k}$$

where $r_1,\ldots,r_k$ are non-negative integers such that $r_1 + \cdots + r_k = n$.

PROOF　　Consider

$$(x_1 + \cdots + x_k)^n = (x_1 + \cdots + x_k)(x_1 + \cdots + x_k)\ldots(x_1 + \cdots + x_k)\,.$$

In expanding this, we get a term $x_1^{r_1} \ldots x_k^{r_k}$ by choosing x_1 from r_1 of the brackets, x_2 from r_2 brackets, and so on. The number of ways of doing this is

$$\binom{n}{r_1,\,\ldots,\,r_k},$$

so this is the coefficient of $x_1^{r_1} \ldots x_k^{r_k}$ in the expansion.　■

Examples

(1) The expansion of $(x + y + z)^3$ is

$$(x+y+z)^3 = x^3+y^3+z^3+3x^2y+3xy^2+3x^2z+3xz^2+3y^2z+3yz^2+6xyz\,.$$

(2) The coefficient of $x^2y^3z^2$ in the expansion of $(x + y + z)^7$ is

$$\binom{7}{2,\,3,\,2} = \frac{7!}{2!\,3!\,2!} = 210\,.$$

(3) Find the coefficient of x^3 in the expansion of $(1 - \frac{1}{x^3} + 2x^2)^5$.

Answer A typical term in this expansion is

$$\binom{5}{a, b, c} \cdot 1^a \cdot (\frac{-1}{x^3})^b \cdot (2x^2)^c$$

where $a + b + c = 5$ (and $a, b, c \geq 0$). To make this a term in x^3, we need

$$-3b + 2c = 3 \quad \text{and } a + b + c = 5 .$$

From the first equation, 3 divides c, so $c = 0$ or 3. If $c = 0$ then $b = -1$, which is impossible. Hence $c = 3$, and it follows that $a = 1, b = 1$. Thus there is just one term in x^3, namely

$$\binom{5}{1, 1, 3}(\frac{-1}{x^3})(2x^2)^3 = -160x^3 .$$

In other words, the coefficient is -160.

Exercises for Chapter 15

1. Liebeck, Einstein and Hawking pinch their jokes from a joke book which contains 12 jokes. Each year Liebeck tells 6 jokes, Einstein tells 4 and Hawking tells 2 (and everyone tells different jokes). For how many years can they go on, never telling the same three sets of jokes?

2. Prove that $\binom{n+1}{r} = \binom{n}{r} + \binom{n}{r-1}$.

3. Three tickets are chosen from a set of 100 tickets numbered $1, 2, 3, \ldots,$ 100. Find the number of choices such that the numbers on the three tickets are

    (a) in arithmetic progression (i.e., $a, a+d, a+2d$ for some a, d)

    (b) in geometric progression (i.e., a, ar, ar^2 for some a, r).

4. The digits $1, 2, 3, 4, 5, 6$ are written down in some order to form a six-digit number.

    (a) How many such six-digit numbers are there altogether?

    (b) How many such numbers are there which are even?

    (c) How many which are divisible by 4?

    (d) How many which are divisible by 8? (*Hint:* First show that the remainder on dividing a six-digit number $abcdef$ by 8 is $4d + 2e + f$.)

5. (a) Find the coefficient of x^{15} in $(1+x)^{18}$.

    (b) Find the coefficient of x^4 in $(2x^3 - \frac{1}{x^2})^8$.

    (c) Find the constant term in the expansion of $(y + x^2 - \frac{1}{xy})^{10}$.

6. The rules of a lottery are as follows: you select 10 numbers between 1 and 50. On lottery night, celebrity critic Ivor Smallbrain chooses at random 6 "correct" numbers. If your 10 numbers include all 6 correct ones, you win.

    Work out your chance of winning the lottery.

7. n points are placed on a circle, and each pair of points is joined by a straight line. The points are chosen so that no three of these lines pass through the same point. Let r_n be the number of regions into which the interior of the circle is divided.

    Draw pictures to calculate r_n for some small values of n.

    Show that for any n,

    $$r_n = n + \binom{n-1}{2} + \binom{n-1}{3} + \binom{n-1}{4}.$$

8. In a lecture I gave the other day, the audience consisted of five mathematicians. Each mathematician fell asleep exactly twice during the lecture. For each pair of mathematicians, there was a moment during the lecture when they were both asleep. Prove that there was a moment when three of the mathematicians were simultaneously asleep.

Chapter 16

More on Sets

In this chapter we develop a little of the theory of sets. Most of the material is rather easy, and much of it is devoted to various definitions and notations to be used in future chapters.

Unions and Intersections

We begin with a couple of definitions.

DEFINITION   *Let A and B be sets. The* union *of A and B, written $A \cup B$, is the set consisting of all elements which lie in either A or B (or both). Symbolically,*

$$A \cup B = \{x \mid x \in A \text{ or } x \in B\}\,.$$

The intersection *of A and B, written $A \cap B$, is the set consisting of all elements which lie in both A and B; thus*

$$A \cap B = \{x \mid x \in A \text{ and } x \in B\}\,.$$

Examples

(1) If $A = \{1, 2, 3\}$ and $B = \{2, 4\}$, then $A \cup B = \{1, 2, 3, 4\}$ and $A \cap B = \{2\}$.

(2) Let $A = \{n \mid n \in \mathbb{Z}, n \geq 0\}$ and $B = \{n \mid n \in \mathbb{Z}, n \leq 0\}$. Then $A \cup B = \mathbb{Z}$ and $A \cap B = \{0\}$.

We say that A and B are *disjoint* sets if they have no elements in common, i.e, if $A \cap B = \emptyset$, the empty set.

Recall from Chapter 1 that the notation $A \subseteq B$ means that A is a subset of B, i.e, every element of A lies in B, which is to say $x \in A \Rightarrow x \in B$. Also, we define $A = B$ to mean that A and B have exactly the same elements. Other ways of expressing $A = B$ are: both $A \subseteq B$ and $B \subseteq A$; or $x \in A \Leftrightarrow x \in B$.

The "algebra of sets" consists of general results involving sets, unions and intersections. Such results are usually pretty uninteresting. Here is one such.

PROPOSITION 16.1
Let A, B, C be sets. Then

$$A \cap (B \cup C) = (A \cap B) \cup (A \cap C).$$

PROOF This just involves keeping careful track of the definitions:

$$x \in A \cap (B \cup C) \Leftrightarrow x \in A \text{ and } x \in (B \text{ or } C)$$
$$\Leftrightarrow (x \in A \text{ and } x \in B) \text{ or } (x \in A \text{ and } x \in C)$$
$$\Leftrightarrow x \in (A \cap B) \cup (A \cap C).$$

Hence, $A \cap (B \cup C) = (A \cap B) \cup (A \cap C)$. ∎

More examples of results in the algebra of sets can be found in the exercises at the end of the chapter.

We can extend the definitions of union and intersection to many sets: if $A_1, A_2, \ldots, A_n$ are sets, their union and intersection are defined as

$$A_1 \cup A_2 \cup \ldots \cup A_n = \{x \mid x \in A_i \text{ for some } i\},$$
$$A_1 \cap A_2 \cap \ldots \cap A_n = \{x \mid x \in A_i \text{ for all } i\}.$$

We sometimes use the more concise notation

$$A_1 \cup \ldots \cup A_n = \bigcup_{i=1}^{n} A_i, \quad A_1 \cap \ldots \cap A_n = \bigcap_{i=1}^{n} A_i.$$

Likewise, if we have an infinite collection of sets $A_1, A_2, A_3, \ldots$, their union and intersection are defined as

$$\bigcup_{i=1}^{\infty} A_i = \{x \mid x \in A_i \text{ for some } i\}, \quad \bigcap_{i=1}^{\infty} A_i = \{x \mid x \in A_i \text{ for all } i\}.$$

Example
For $i \geq 1$ let $A_i = \{x \mid x \in \mathbb{Z}, x \geq i\}$. Then

$$\bigcup_{i=1}^{\infty} A_i = \mathbb{N}, \quad \bigcap_{i=1}^{\infty} A_i = \emptyset.$$

If A, B are sets, their *difference* is defined to be the set

$$A - B = \{x \mid x \in A \text{ and } x \notin B\} \, .$$

For example, if $A = \{x \mid x \in \mathbb{R}, 0 \le x \le 1\}$ and $B = \mathbb{Q}$, then $A - B$ is the set of irrationals between 0 and 1.

Cartesian Products

Cartesian products give a way of constructing new sets from old.

DEFINITION   *Let A, B be sets. The* Cartesian product *of A and B, written $A \times B$, is the set consisting of all symbols of the form (a, b) with $a \in A$, $b \in B$. Such a symbol (a, b) is called an* ordered pair *of elements of A and B. Two ordered pairs (a, b), (a', b') are deemed to be equal if and only if both $a = a'$ and $b = b'$.*

For example, if $A = \{1, 2\}$ and $B = \{1, 4, 5\}$, then $A \times B$ consists of the six ordered pairs

$$(1, 1), (1, 4), (1, 5), (2, 1), (2, 4), (2, 5) \, .$$

As another example, when $A = B = \mathbb{R}$, the Cartesian product is $\mathbb{R} \times \mathbb{R}$, which consists of all ordered pairs (x, y) $(x, y \in \mathbb{R})$, commonly known as coordinate pairs of points in the plane.

We can also form the Cartesian product of more than two sets in a similar way: if $A_1, A_2, \ldots, A_n$ are sets, their Cartesian product is defined to be the set $A_1 \times A_2 \times \ldots \times A_n$ consisting of all symbols of the form $(a_1, a_2, \ldots, a_n)$, where $a_i \in A_i$ for all i. Such symbols are called *n-tuples* of elements of $A_1, \ldots, A_n$.

Finite Sets

Logically enough, we call a set S a *finite* set if it has only a finite number of elements. If S has n elements, we write $|S| = n$.

If a set is not finite, it is said to be an *infinite* set.

For example, if $S = \{1, -3, \sqrt{2}\}$, then S is finite and $|S| = 3$. And $\mathbb{Z}$ is an infinite set.

Here is a useful result about finite sets.

PROPOSITION 16.2
If A and B are finite sets, then

$$|A \cup B| = |A| + |B| - |A \cap B|.$$

PROOF   Let $|A \cap B| = k$, say $A \cap B = \{x_1, \ldots, x_k\}$. These elements, and no others, belong to both A and B, so we can write

$$A = \{x_1, \ldots, x_k, a_1, \ldots, a_l\}, \qquad B = \{x_1, \ldots, x_k, b_1, \ldots, b_m\},$$

where $|A| = k + l$, $|B| = k + m$. Then

$$A \cup B = \{x_1, \ldots, x_k, a_1, \ldots, a_l, b_1, \ldots, b_m\},$$

and all these elements are different, so

$$|A \cup B| = k + l + m = (k + l) + (k + m) - k$$
$$= |A| + |B| - |A \cap B|. \quad \blacksquare$$

Example
Out of a total of 30 students, 19 are doing mathematics, 17 are doing music, and 10 are doing both. How many are doing neither?

Answer Let A be the set doing mathematics and B the set doing music. Then

$$|A| = 19, \ |B| = 17, \ |A \cap B| = 10.$$

Hence, Proposition 16.2 gives $|A \cup B| = 19 + 17 - 10 = 26$. Since there are 30 students in all, there are therefore 4 doing neither mathematics nor music.

A result similar to Proposition 16.2, for the union of three sets, is true: if A, B, C are finite sets then

$$|A \cup B \cup C| = |A| + |B| + |C| - |A \cap B| - |A \cap C| - |B \cap C| + |A \cap B \cap C|.$$

This is set as Exercise 2 at the end of the chapter. In fact, this result and Proposition 16.2 are special cases of a general result, known as the "Inclusion-Exclusion Principle," which says that if $A_1, A_2, \ldots, A_n$ are finite sets, then the size of the union $A_1 \cup A_2 \cup \ldots \cup A_n$ can be calculated by adding together the sizes of all intersections of an odd number of these sets, and subtracting from this the sizes of all intersections of an even number of the sets. This is not a terribly difficult result, and can be proved using induction, but we shall not do this here.

Example

How many integers are there between 1000 and 9999 which contain the digits 0, 8 and 9 at least once each? (For example, 8950 and 8089 are such integers.)

Answer Let $S = \{1000, 1001, \ldots, 9999\}$. For $k = 0, 8$ or 9, let A_k be the set of integers in S which have no digit equal to k. Then $A_0 \cup A_8 \cup A_9$ is the set of integers in S which are missing either 0, 8 or 9, and so the number we are asked for in the question is

$$|S| - |A_0 \cup A_8 \cup A_9| = 9000 - |A_0 \cup A_8 \cup A_9|.$$

We therefore need to calculate $|A_0 \cup A_8 \cup A_9|$, which we shall do using the above equality for $|A \cup B \cup C|$. By the Multiplication Principle 15.1, we have $|A_9| = 8 \times 9 \times 9 \times 9$, since there are 8 choices for the first digit (it cannot be 0 or 9), and 9 for each of the others. Similarly,

$$|A_8| = 8 \times 9 \times 9 \times 9 = 5832, \quad |A_0| = 9 \times 9 \times 9 \times 9 = 6561,$$
$$|A_0 \cap A_8| = |A_0 \cap A_9| = 8 \times 8 \times 8 \times 8 = 4096,$$
$$|A_8 \cap A_9| = 7 \times 8 \times 8 \times 8 = 3584,$$
$$|A_0 \cap A_8 \cap A_9| = 7 \times 7 \times 7 \times 7 = 2401.$$

Therefore by the above equality for $|A \cup B \cup C|$,

$$|A_0 \cup A_8 \cup A_9| = 5832 + 5832 + 6561 - 4096 - 4096 - 3584 + 2401 = 8850.$$

Hence, the number of integers in S which have at least one of each of the digits 0, 8 and 9 is equal to $9000 - 8850 = 150$.

Exercises for Chapter 16

1. (a) Let A, B be sets. Prove that $A \cup B = A$ if and only if $B \subseteq A$.

   (b) Prove that $(A - C) \cap (B - C) = (A \cap B) - C$ for all sets A, B, C.

2. Prove that if A, B, C are finite sets, then

$$|A \cup B \cup C| = |A| + |B| + |C| - |A \cap B| - |A \cap C| - |B \cap C| + |A \cap B \cap C|.$$

3. (a) 73% of British people like cheese, 76% like apples and 10% like neither. What percentage like both cheese and apples?

   (b) In a class of 30 children, everyone supports at least one of three teams: 16 support Manchester United, 17 support Stoke City, and 14 support Doncaster Rovers; also 8 support both United and City, 7 both United and Rovers, and 9 both City and Rovers. How many support all three teams?

4. (a) Find the number of integers between 1 and 5000 which are divisible by neither 3 nor 4.

   (b) Find the number of integers between 1 and 5000 which are divisible by neither 3 nor 4 nor 5.

   (c) Find the number of integers between 1 and 5000 which are divisible by one or more of the numbers 4, 5 and 6.

5. Let S be a finite set consisting of n elements. Show that the total number of subsets of S is 2^n.

6. Critic Ivor Smallbrain is travelling on a transatlantic plane. The on-board film is *Everything You Wanted to Know About Sets But Were Afraid to Ask,* and Ivor is bored. So he decides to ask every passenger their nationality. Having done this, he works out that on the plane there are 9 boys, 5 American children, 9 men, 7 non-American boys, 14 Americans, 6 American males, and 7 non-American females. How many people are there on the plane altogether?

Chapter 17

Equivalence Relations

Let S be a set. A *relation* on S is defined as follows. We choose a subset R of the Cartesian product $S \times S$; in other words, R consists of some of the ordered pairs (s, t) with $s, t \in S$. For those ordered pairs $(s, t) \in R$, we write $s \sim t$ and say s is *related* to t. And for $(s, t) \notin R$, we write $s \not\sim t$. Thus, the symbol $\sim$ relates various pairs of elements of S. It is called a *relation* on S.

This definition probably seems a bit strange at first sight. A few examples should serve to clarify matters.

Example 17.1

Here are seven examples of relations on various sets S.

(1) Let $S = \mathbb{R}$, and define $a \sim b \Leftrightarrow a < b$. Here $R = \{(s, t) \in \mathbb{R} \times \mathbb{R} \mid s < t\}$.

(2) Let $S = \mathbb{Z}$ and let m be a positive integer. Define $a \sim b \Leftrightarrow a \equiv b \bmod m$.

(3) $S = \mathbb{C}$, and $a \sim b \Leftrightarrow |a - b| < 1$.

(4) $S = \{1, 2\}$, and $\sim$ defined by $1 \sim 1$, $1 \sim 2$, $2 \not\sim 1$, $2 \sim 2$.

(5) $S = \{1, 2\}$, and $\sim$ defined by $1 \sim 1$, $1 \not\sim 2$, $2 \not\sim 1$, $2 \sim 2$.

(6) $S =$ all people in Britain, and $a \sim b$ if and only if a and b have the same father.

(7) S any set, and $a \sim b \Leftrightarrow a = b$.

The relations on a set S correspond to the subsets of $S \times S$, and there is nothing much more to say about them in general. However, there are certain types of relations that are worthy of study, as they crop up frequently all over the place. These are called *equivalence relations*. Here is the definition.

DEFINITION   *Let S be a set, and let $\sim$ be a relation on S. Then $\sim$ is an equivalence relation if the following three properties hold for all $a, b, c \in S$:*

(i) $a \sim a$ *(this says $\sim$ is* reflexive*)*

(ii) if $a \sim b$ then $b \sim a$ *(this says $\sim$ is* symmetric*)*

(iii) if $a \sim b$ and $b \sim c$ then $a \sim c$ *(this says $\sim$ is* transitive*)*

Let us examine each of the Examples 17.1 for these properties.

In Example 17.1(1), $S = \mathbb{R}$ and $a \sim b \Leftrightarrow a < b$. This is not reflexive or symmetric, but it is transitive (as $a < b$ and $b < c \Rightarrow a < c$).

Example 17.1(2) is an equivalence relation by Proposition 14.2.

Now consider Example 17.1(3), where $S = \mathbb{C}$ and $a \sim b \Leftrightarrow |a - b| < 1$. This is reflexive and symmetric. But it is not transitive; to see this, take $a = \frac{3}{4}, b = 0, c = -\frac{3}{4}$: then $|a - b| < 1$, $|b - c| < 1$ but $|a - c| > 1$.

The relation in Example 17.1(4) is reflexive and transitive, but not symmetric; and the relation in Example 17.1(5) is an equivalence relation.

I leave it to you to show that the relations in Examples 17.1(6) and (7) are both equivalence relations.

Equivalence Classes

Let S be a set and $\sim$ an equivalence relation on S. For $a \in S$, define

$$cl(a) = \{s \mid s \in S, s \sim a\} .$$

Thus, $cl(a)$ is the set of things which are related to a. The subset $cl(a)$ is called an *equivalence class* of $\sim$. The equivalence classes of $\sim$ are the subsets $cl(a)$ as a ranges over the elements of S.

Example 17.2

Let m be a positive integer, and let $\sim$ be the equivalence relation on $\mathbb{Z}$ defined as in Example 17.1(2) — that is,

$$a \sim b \Leftrightarrow a \equiv b \bmod m .$$

What are the equivalence classes of this relation?

To answer this, let us write down various equivalence classes:

$$cl(0) = \{s \in \mathbb{Z} \mid s \equiv 0 \bmod m\} ,$$
$$cl(1) = \{s \in \mathbb{Z} \mid s \equiv 1 \bmod m\} , \ldots$$
$$cl(m - 1) = \{s \in \mathbb{Z} \mid s \equiv m - 1 \bmod m\} .$$

We claim that these are *all* the equivalence classes. For if n is any integer, then by Proposition 11.1 there are integers q, r such that $n = qm + r$ with $0 \leq r < m$. Then $n \equiv r \bmod m$, so $n \in cl(r)$ which is one of the classes listed above; and moreover,

$$cl(n) = \{s \in \mathbb{Z} \mid s \equiv n \bmod m\} = \{s \in \mathbb{Z} \mid s \equiv r \bmod m\} = cl(r) .$$

Hence, any equivalence class $cl(n)$ is equal to one of those listed above.

We conclude that in this example, there are exactly m different equivalence classes, $cl(0), cl(1), \ldots, cl(m-1)$. Note also that every integer lies in exactly one of these classes.

Example 17.3

Consider now the equivalence relation defined in Example 17.1(6): $S =$ all people in Britain, and $a \sim b$ if and only if a and b have the same father. What are the equivalence classes?

If $a \in S$, then $cl(a)$ is the set of all people with the same father as a. In other words, if f is the father of a then $cl(a)$ consists of all the children of f. So one way of listing all the equivalence classes is as follows: let $f_1, \ldots, f_n$ be a list of all fathers of people in Britain; if C_i is the set of children of f_i living in Britain, then the equivalence classes are $C_1, \ldots, C_n$.

We now prove a general property of equivalence classes. Recall from Chapter 15 that a *partition* of a set S is a collection of subsets $S_1, \ldots, S_k$ such that each element of S lies in exactly one of these subsets. Another way of putting this is that the subsets $S_1, \ldots, S_k$ have the properties that their union is S, and any two of them are disjoint (i.e., $S_i \cap S_j = \emptyset$ for any $i \neq j$).

For example, if $S = \{1, 2, 3, 4, 5\}$ then the subsets $\{1\}, \{2, 4\}, \{3, 5\}$ form a partition of S, whereas the subsets $\{1\}, \{2, 4\}, \{3\}, \{4, 5\}$ do not.

PROPOSITION 17.1

Let S be a set and let $\sim$ be an equivalence relation on S. Then the equivalence classes of $\sim$ form a partition of S.

PROOF   If $a \in S$, then since $a \sim a$, a lies in the equivalence class $cl(a)$.

We need to show that a lies in only one equivalence class. So suppose that a lies in $cl(s)$ and $cl(t)$; in other words, $a \sim s$ and $a \sim t$. We show that this implies that $cl(s) = cl(t)$.

Let $x \in cl(s)$. Then $x \sim s$. Also $s \sim a$ and $a \sim t$, so by transitivity, $x \sim t$. Hence

$$x \in cl(s) \Rightarrow x \in cl(t) \, .$$

Similarly, if $x \in cl(t)$ then $x \sim t$, and also $t \sim a$ and $a \sim s$, so $x \sim s$, showing

$$x \in cl(t) \Rightarrow x \in cl(s) \, .$$

We conclude that $cl(s) = cl(t)$, as required. Thus, any element a of S lies in exactly one equivalence class.   ∎

Exercises for Chapter 17

1. Which of the following relations are equivalence relations on the given set S?

   (i) $S = \mathbb{R}$, and $a \sim b \Leftrightarrow a = b$ or $-b$.

   (ii) $S = \mathbb{Z}$, and $a \sim b \Leftrightarrow ab = 0$.

   (iii) $S = \mathbb{R}$, and $a \sim b \Leftrightarrow a^2 + a = b^2 + b$.

   (iv) S is the set of all people in the world, and $a \sim b$ means a lives within 100 miles of b.

   (v) S is the set of all points in the plane, $a \sim b$ means a and b are the same distance from the origin.

   (vi) $S = \mathbb{N}$, and $a \sim b \Leftrightarrow ab$ is a square.

   (vii) $S = \{1, 2, 3\}$, and $a \sim b \Leftrightarrow a = 1$ or $b = 1$.

   (viii) $S = \mathbb{R} \times \mathbb{R}$, and $(x, y) \sim (a, b) \Leftrightarrow x^2 + y^2 = a^2 + b^2$.

2. For those relations in Question 1 which are equivalence relations, describe the equivalence classes.

3. Let $S = \{1, 2, 3, 4\}$, and suppose that $\sim$ is an equivalence relation on S. You are given the information that $1 \sim 2$ and $2 \sim 3$.

   Show that there are exactly two possibilities for the relation $\sim$, and describe both (i.e., for all $a, b \in S$ say whether or not $a \sim b$).

4. Critic Ivor Smallbrain is sitting through a showing of the latest Disney film *101 Equivalence Relations*. Ivor is fed up, and starts to discuss with his colleague Luigi Paparazzi how many different equivalence relations they can find on the set $\{1, 2\}$. They find just 2. Then on the set $\{1, 2, 3\}$ they find just 5 different equivalence relations.

   Have they found *all* the equivalence relations on these sets? How many should they find on $\{1, 2, 3, 4\}$ and on $\{1, 2, 3, 4, 5\}$? Investigate further if you feel like it!

Chapter 18

Functions

Much of mathematics and its applications is concerned with the study of functions of various kinds. In this chapter we give the definition and some elementary examples, and introduce certain important general types of functions.

DEFINITION   *Let S and T be sets. A function from S to T is a rule which assigns to each $s \in S$ a single element of T, denoted by $f(s)$. We write*

$$f : S \rightarrow T$$

to mean that f is a function from S to T. If $f(s) = t$, we often say f sends $s \rightarrow t$.

If $f : S \rightarrow T$ is a function, the image *of f is the set of all elements of T which are equal to $f(s)$ for some $s \in S$. We write $f(S)$ for the image of f. Thus*

$$f(S) = \{ f(s) \,|\, s \in S \} \,.$$

Examples 18.1

(1) Define $f : \{1, 2, 3\} \rightarrow \mathbb{Z}$ by $f(x) = x^2 - 4$ for $x \in \{1, 2, 3\}$. The image of f is $\{-3, 0, 5\}$.

(2) Define $f : \mathbb{R} \rightarrow \mathbb{R}$ by $f(x) = x^2$ for all $x \in \mathbb{R}$. The image of f is $f(\mathbb{R}) = \{ y \,|\, y \in \mathbb{R}, y \geq 0 \}$.

(3) A body is dropped and falls under gravity for 1 second. The distance travelled at time t is $\frac{1}{2}gt^2$. If we call this distance $s(t)$, and write $I = \{t \in \mathbb{R} \,|\, 0 \leq t \leq 1\}$, then s is a function from I to $\mathbb{R}$ defined by $s(t) = \frac{1}{2}gt^2$. The image of s is the set of reals between 0 and $\frac{g}{2}$.

(4) Define $f : \mathbb{N} \times \mathbb{N} \rightarrow \mathbb{Z}$ by $f(m, n) = m - n$ for all $m, n \in \mathbb{N}$. The image of f is $\mathbb{Z}$.

(5) Let $S = \{a, b, c\}$ and define functions $f : S \rightarrow S$ and $g : S \rightarrow S$ as follows:

$$f \text{ sends } a \rightarrow b, \ b \rightarrow c, \ c \rightarrow a, \quad g \text{ sends } a \rightarrow b, \ b \rightarrow c, \ c \rightarrow b \,.$$

Then $f(S) = S$, while $g(S) = \{b, c\}$.

(6) Let S be any set, and define a function $\iota_S : S \to S$ by

$$\iota_S(s) = s \quad \text{for all } s \in S.$$

This function ι_S is called the *identity* function of S.

We now define certain important types of functions.

DEFINITION   Let $f : S \to T$ be a function.
(I) We say f is onto if the image $f(S) = T$, i.e., if for every $t \in T$ there exists $s \in S$ such that $f(s) = t$.
(II) We say f is one-to-one (usually written simply as 1-1) if whenever $s_1, s_2 \in S$ with $s_1 \neq s_2$, then $f(s_1) \neq f(s_2)$; in other words, f is 1-1 if f sends different elements of S to different elements of T. Another way of putting this is to say that for all $s_1, s_2 \in S$,

$$f(s_1) = f(s_2) \Rightarrow s_1 = s_2.$$

This is usually the most useful definition to use when testing whether functions are 1-1.
(III) We say f is a bijection if f is both onto and 1-1.

Note Functions that are onto are often called *surjective* functions, or *surjections*; and functions that are 1-1 are often called *injective* functions or *injections*. You will find these terms in many books, but I prefer to stick to the slightly more descriptive terms "onto" and "1-1."

Let us briefly discuss which of these properties the functions in Examples 18.1 possess.

The function in Example 18.1(1) sends $1 \to -3$, $2 \to 0$, $3 \to 5$, so it is 1-1. It is clearly not onto.

The function in Example 18.1(2) is not onto, and is not 1-1 either, since it sends 1 and -1 to the same thing.

On the other hand, the function $s : I \to \mathbb{R}$ in Example 18.1(3) is 1-1, since for $t_1, t_2 \in I$,

$$s(t_1) = s(t_2) \Rightarrow \frac{1}{2}gt_1^2 = \frac{1}{2}gt_2^2 \Rightarrow t_1 = t_2.$$

Also s is not onto.

The function in Example 18.1(4) is onto, but is not 1-1 since, for example, it sends both $(1, 1)$ and $(2, 2)$ to 0.

In Example 18.1(5), the function f is a bijection, while g is neither 1-1 nor onto. Finally, the identity function in Example 18.1(6) is a bijection.

Here is a quite useful result relating 1-1 and onto functions to the sizes of sets.

PROPOSITION 18.1

Let $f : S \to T$ be a function, where S and T are finite sets.
(i) *If f is onto, then $|S| \geq |T|$.*
(ii) *If f is 1-1, then $|S| \leq |T|$.*
(iii) *If f is a bijection, then $|S| = |T|$.*

PROOF   (i) Let $|S| = n$ and write $S = \{s_1, \ldots, s_n\}$. As f is onto, we have

$$T = f(S) = \{f(s_1), \ldots, f(s_n)\} \ .$$

Hence $|T| \leq n$. (Of course $|T|$ could be less than n, as some of the $f(s_i)$s could be equal.)

(ii) Again let $|S| = n$ and $S = \{s_1, \ldots, s_n\}$. As f is 1-1, the elements $f(s_1), \ldots, f(s_n)$ are all different, and lie in T. Therefore $|T| \geq n$.

(iii) If f is a bijection, then $|S| \geq |T|$ by (i), and $|S| \leq |T|$ by (ii), so $|S| = |T|$.   ∎

Part (ii) of Proposition 18.1 implies that if $|S| > |T|$ then there is no 1-1 function from S to T. This can be phrased somewhat more strikingly in the following way:

If we put $n + 1$ or more pigeons into n pigeonholes, then there must be a pigeonhole containing more than one pigeon.

(For if no pigeonhole contained more than one pigeon, the function sending pigeons to their pigeonholes would be 1-1.)

The above statement is known as the *pigeonhole principle*, and it is surprisingly useful. As a very simple example, in any group of 13 or more people, at least two must have their birthday in the same month (here the people are the "pigeons," and the 12 months are the "pigeonholes"). As another example, in any set of 6 integers, there must be two whose difference is divisible by 5: to see this, regard the 6 integers as the pigeons, and their remainders on division by 5 as the pigeonholes. More examples of the use of the pigeonhole principle can be found in Exercise 4 at the end of the chapter.

Inverse Functions

Given a function $f : S \to T$, under what circumstances can we define an "inverse function" from T to S, sending everything back to where it came from? (In other words, if f sends $s \to t$, the "inverse" function should send $t \to s$.) To define such a function from T to S, we need

(a) f to be onto (otherwise some elements of T will not be sent anywhere by the inverse function), and

(b) f to be 1-1 (otherwise some element of T will be sent back to more than one element of S).

In other words, to be able to define such an inverse function from T to S, we need f to be a bijection. Here is the formal definition.

DEFINITION   *Let $f : S \to T$ be a bijection. The* inverse *function of f is the function from $T \to S$ which sends each $t \in T$ to the unique $s \in S$ such that $f(s) = t$. We denote the inverse function by $f^{-1} : T \to S$. Thus for $s \in S, t \in T$,*

$$f^{-1}(t) = s \Leftrightarrow f(s) = t .$$

As a consequence we have

$$f^{-1}(f(s)) = s \quad and \quad f(f^{-1}(t)) = t$$

for all $s \in S, t \in T$.

Examples

(1) Let $S = \{a, b, c\}$ and let $f : S \to S$ be the function which sends $a \to b$, $b \to c$, $c \to a$. Then f is a bijection, and the inverse function $f^{-1} : S \to S$ sends everything back to where it came from, namely $a \to c$, $b \to a$, $c \to b$.

(2) Define $f : \mathbb{R} \to \mathbb{R}$ by $f(x) = 8 - 2x$ for all $x \in \mathbb{R}$. Then f is a bijection, and $f^{-1}(t) = \frac{1}{2}(8 - t)$ for all $t \in \mathbb{R}$.

Composition of Functions

Composition gives us a useful way of combining two functions to form another one. Here is the definition.

DEFINITION   *Let S, T, U be sets, and let $f : S \to T$ and $g : T \to U$ be functions. The* composition *of f and g is the function $g \circ f : S \to U$ which is defined by the rule*

$$(g \circ f)(s) = g(f(s)) \quad for\ all\ s \in S .$$

(Thus $g \circ f$ is just a "function of a function," which is a phrase you may have seen before.)

Examples

(1) Let $f : \mathbb{R} \to \mathbb{R}, \quad g : \mathbb{R} \to \mathbb{R}$ be defined by

$$f(x) = \sin x, \quad g(x) = x^2 + 1$$

for all $x \in \mathbb{R}$. Then both compositions $g \circ f$ and $f \circ g$ are functions from $\mathbb{R} \to \mathbb{R}$, and

$$g \circ f(x) = g(f(x)) = \sin^2 x + 1, \quad f \circ g(x) = \sin\left(x^2 + 1\right)$$

for all $x \in \mathbb{R}$.

(2) Let $f : \{1, 2, 3\} \to \mathbb{Z}$ and $g : \mathbb{Z} \to \mathbb{N}$ be defined by

$$f \text{ sends } 1 \to 0, \ 2 \to -5, \ 3 \to 7, \quad \text{and}$$
$$g(x) = |x| \quad \text{for all } x \in \mathbb{Z}.$$

Then $g \circ f : \{1, 2, 3\} \to \mathbb{N}$ sends $1 \to 0, \ 2 \to 5, \ 3 \to 7$, and $f \circ g$ does not exist.

Notice that if $f : S \to T$ is a bijection, then by definition of the inverse function $f^{-1} : T \to S$, we have

$$\left(f^{-1} \circ f\right)(s) = s, \quad \left(f \circ f^{-1}\right)(t) = t$$

for all $s \in S, t \in T$. Another way of putting this is to say that

$$f^{-1} \circ f = \iota_S, \quad f \circ f^{-1} = \iota_T,$$

where ι_S, ι_T are the identity functions of S and T, as defined in Example 18.1(6).

Here is a neat result linking composition with the properties of being 1-1 or onto.

PROPOSITION 18.2

Let S, T, U be sets, and let $f : S \to T$ and $g : T \to U$ be functions. Then
(a) *if f and g are both 1-1, so is $g \circ f$,*
(b) *if f and g are both onto, so is $g \circ f$,*
(c) *if f and g are both bijections, so is $g \circ f$.*

PROOF   (a) If f, g are both 1-1, then for $s_1, s_2 \in S$,

$$(g \circ f)(s_1) = (g \circ f)(s_2) \Rightarrow g(f(s_1)) = g(f(s_2))$$
$$\Rightarrow f(s_1) = f(s_2) \quad \text{as } g \text{ is 1-1}$$
$$\Rightarrow s_1 = s_2 \quad \text{as } f \text{ is 1-1}$$

and hence $g \circ f$ is 1-1.

(b) Suppose f, g are both onto. For any $u \in U$, there exists $t \in T$ such that $g(t) = u$ (as g is onto), and there exists $s \in S$ such that $f(s) = t$ (as f is onto). Hence $(g \circ f)(s) = g(f(s)) = g(t) = u$, showing that $g \circ f$ is onto.

(c) This follows immediately from parts (a) and (b). ∎

Counting Functions

How many functions are there from one finite set to another? This question is quite easily answered using some of our counting methods from Chapter 15:

PROPOSITION 18.3
Let S, T be finite sets, with $|S| = m$, $|T| = n$. Then the number of functions from S to T is equal to n^m.

PROOF   Let $S = \{s_1, s_2, \ldots, s_m\}$. Defining a function $f : S \to T$ is an m-stage process:

Stage 1 : choose $f(s_1)$; this can be any of the n members of T, so the number of choices is n

Stage 2 : choose $f(s_2)$; again, the number of choices is n

And so on, up to

Stage m : choose $f(s_m)$; again, the number of choices is n.

Thus by the Multiplication Principle 15.1, the total number of functions is $n.n. \ldots n = n^m$. ∎

One can also get a formula for the number of 1-1 functions from S to T (see Exercise 6 below).

Exercises for Chapter 18

1. For each of the following functions f, say whether f is $1-1$ and whether f is onto:

   (i) $f : \mathbb{R} \to \mathbb{R}$ defined by $f(x) = x^2 + 2x$ for all $x \in \mathbb{R}$.

   (ii) $f : \mathbb{R} \to \mathbb{R}$ defined by

$$f(x) = x - 2 \quad \text{if } x > 1$$
$$-x \quad \text{if } -1 \leq x \leq 1$$
$$x + 2 \quad \text{if } x < -1 .$$

   (iii) $f : \mathbb{N} \times \mathbb{N} \times \mathbb{N} \to \mathbb{N}$ defined by $f(m, n, r) = 2^m 3^n 5^r$ for all $m, n, r \in \mathbb{N}$.

   (iv) $f : \mathbb{N} \times \mathbb{N} \times \mathbb{N} \to \mathbb{N}$ defined by $f(m, n, r) = 2^m 3^n 6^r$ for all $m, n, r \in \mathbb{N}$.

   (v) Let $\sim$ be the equivalence relation on $\mathbb{Z}$ defined by $a \sim b \Leftrightarrow a \equiv b \bmod 7$, and let S be the set of equivalence classes of $\sim$. Define $f : S \to S$ by $f(\mathrm{cl}(s)) = \mathrm{cl}(s + 1)$ for all $s \in \mathbb{Z}$.

2. The functions $f, g : \mathbb{R} \to \mathbb{R}$ are defined as follows:

$$f(x) = 2x \quad \text{if } 0 \leq x \leq 1, \quad \text{and } f(x) = 1 \text{ otherwise ;}$$
$$g(x) = x^2 \quad \text{if } 0 \leq x \leq 1, \quad \text{and } g(x) = 0 \text{ otherwise .}$$

   Give formulae describing the functions $g \circ f$ and $f \circ g$. Draw the graphs of these functions.

3. Two functions $f, g : \mathbb{R} \to \mathbb{R}$ are such that for all $x \in \mathbb{R}$,

$$g(x) = x^2 + x + 3, \quad \text{and } (g \circ f)(x) = x^2 - 3x + 5 .$$

   Find the possibilities for f.

4. Use the pigeonhole principle to prove the following statements involving a positive integer n:

   (a) In any set of $n + 1$ integers, there must be two whose difference is divisible by n.

   (b) Given any n integers $a_1, a_2, \ldots, a_n$, there is a non-empty subset of these whose sum is divisible by n. (*Hint:* consider the integers $0, a_1, a_1 + a_2, \ldots, a_1 + \cdots + a_n$.)

   (c) Given any set S consisting of ten distinct integers between 1 and 50, there are two different 5-element subsets of S with the same sum.

(d) Given any set T consisting of nine distinct integers between 1 and 50, there are two disjoint subsets of T with the same sum.

5. (a) Find an onto function from $\mathbb{N}$ to $\mathbb{Z}$.

   (b) Find a 1-1 function from $\mathbb{Z}$ to $\mathbb{N}$.

6. (a) If $S = \{1, 2, 3\}$ and $T = \{1, 2, 3, 4, 5\}$, calculate the number of 1-1 functions $f : S \to T$.

   (b) Let $|S| = m$, $|T| = n$ with $m \le n$. Find a formula for the number of 1-1 functions $f : S \to T$.

7. Let a, b, c be three different integers. Prove that it is impossible to find a polynomial $P(x)$ with integer coefficients such that

$$P(a) = b, \quad P(b) = c \quad \text{and } P(c) = a \ .$$

(*Hint:* Observe that $P(x) - P(y) = (x - y)Q(x, y)$ where $Q(x, y)$ is a polynomial in x, y with integer coefficients. Substitute $x = a$, $y = b$, etc. into this equation and see what happens.)

Chapter 19

Infinity

Given two finite sets, it is simple to compare their sizes. For example, we would say that the set of corners of a pentagon is larger than the set of players in a string quartet, simply because the first set has 5 elements, while the second has only 4.

But can we compare the sizes of *infinite* sets in any meaningful way? We have encountered many different infinite sets at various points in this book, such as $\mathbb{N}$, $\mathbb{Z}$, $\mathbb{Q}$, $\mathbb{R}$, $\mathbb{C}$, $\mathbb{N} \times \mathbb{N}$, $\mathbb{Q} \times \mathbb{R} \times \mathbb{C}$, and so on. How can we compare these with each other?

There is a way to do this using functions. To set this up, let us begin with an elementary observation about finite sets. If S is a set of size n, say $S = \{s_1, s_2, \ldots, s_n\}$, then the function $f : S \to \{1, 2, \ldots, n\}$ defined by

$$f(s_1) = 1, \ f(s_2) = 2, \ldots, f(s_n) = n$$

is a bijection. Thus we can say

$$S \text{ has size } n \quad \Leftrightarrow \quad \text{there is a bijection from } S \text{ to } \{1, 2, \ldots, n\}.$$

We now extend this notion to arbitrary sets.

DEFINITION   *Two sets A and B are said to be* equivalent *to each other if there is a bijection from A to B. We write* $A \sim B$ *if A and B are equivalent to each other.*

In accordance with the preamble to the definition, we can informally think of two sets which are equivalent to each other as "having the same size."

Before doing anything else, let us establish that the relation $\sim$ is an equivalence relation on sets.

PROPOSITION 19.1
The relation $\sim$ defined above is an equivalence relation.

PROOF   First we show $\sim$ is reflexive, i.e., $A \sim A$ for any set A. This is true since the identity function $\iota_A : A \to A$ defined by $i_A(a) = a$ for all $a \in A$, is a bijection.

Next we show $\sim$ is symmetric. Suppose $A \sim B$, so there is a bijection $f : A \to B$. Then the inverse function $f^{-1} : B \to A$ is a bijection, so $B \sim A$.

Finally, we show $\sim$ is transitive. Suppose $A \sim B$ and $B \sim C$, so there are bijections $f : A \to B$ and $g : B \to C$. Then by Proposition 18.2, $g \circ f : A \to C$ is a bijection, so $A \sim C$. Hence $\sim$ is transitive.   ∎

Examples

(1) Let $A = \mathbb{N}$ and let $B = \{2n \mid n \in \mathbb{N}\}$, the set of all positive even numbers. Then the function $f : A \to B$ defined by

$$f(n) = 2n \quad \text{for all } n \in \mathbb{N}$$

is a bijection. Thus $A \sim B$, i.e., $\mathbb{N} \sim$ even numbers in $\mathbb{N}$.

This example shows that $\mathbb{N}$ can be equivalent to a subset of itself. (Informally, $\mathbb{N}$ "has the same size" as a subset of itself.) It is of course not possible for any *finite* set to have this property.

(2) Suppose A is a set that is equivalent to $\mathbb{N}$. This means there is a bijection $f : \mathbb{N} \to A$. For $n \in \mathbb{N}$, let $f(n) = a_n \in A$. Since f is onto, we then have

$$A = \{a_1, a_2, a_3, \ldots, a_n, \ldots\} .$$

In other words, we can *list* all the elements of A as $a_1, a_2, a_3, \ldots$.

Countable Sets

The listing property of the last example is so fundamental that we give it a special definition:

DEFINITION   *A set A is said to be* countable *if A is equivalent to $\mathbb{N}$. In other words, A is countable if it is an infinite set, all of whose elements can be listed as $A = \{a_1, a_2, a_3, \ldots, a_n, \ldots\}$.*

Examples

(1) $\mathbb{N}$ is obviously itself countable: its elements can be listed as $1, 2, 3, \ldots$.

(2) The set $B = \{2n \mid n \in \mathbb{N}\}$ of positive even numbers is countable: the elements of B can be listed a $2, 4, 6, 8, \ldots$.

(3) What about $\mathbb{Z}$ — is it countable? This is not quite so obvious, but the answer is yes because we can list all the elements of $\mathbb{Z}$ as $0, 1, -1, 2, -2, 3, -3, \ldots$. Correspondingly, we could define a bijection $f : \mathbb{N} \to \mathbb{Z}$ by

$$f(2n) = n \quad \text{and} \quad f(2n-1) = -(n-1)$$

for all $n \geq 1$.

The same idea shows that the union of any two countable sets is countable (just list the elements alternately, omitting any repetitions).

The next proposition provides us with many more examples of countable sets.

PROPOSITION 19.2

Every infinite subset of $\mathbb{N}$ is countable.

PROOF Let S be an infinite subset of $\mathbb{N}$. Take s_1 to be the smallest integer in S; then take s_2 to be the smallest integer in $S - \{s_1\}$, take s_3 to be the smallest in $S - \{s_1, s_2\}$, and so on. In this way, we list all the elements of S in ascending order as $S = \{s_1, s_2, s_3, \ldots\}$. Therefore, S is countable. (A bijection $f : \mathbb{N} \to S$ would be simply $f(1) = s_1$, $f(2) = s_2$, $f(3) = s_3$, and so on.) ∎

Let us now consider the question of whether $\mathbb{Q}$, the set of rationals, is countable. This is much more subtle than any of the above examples. For a start, we certainly cannot list the positive rationals in ascending order, since whatever rational x we started the list with, there would be a smaller one (e.g., $\frac{1}{2}x$) which then would not appear on the list. However, could it be possible to devise a devilishly clever alternative way to list the rationals?

Somewhat amazingly, the answer is yes:

PROPOSITION 19.3

The set of rationals $\mathbb{Q}$ is countable.

PROOF First consider $\mathbb{Q}^+$, the set of positive rationals. We show how to list the elements of $\mathbb{Q}^+$. The key is first to write the positive rationals in an array as follows:

$$
\begin{array}{ccccccc}
\frac{1}{1} & \frac{2}{1} & \frac{3}{1} & \frac{4}{1} & \frac{5}{1} & \cdots \\[4pt]
\frac{1}{2} & \frac{2}{2} & \frac{3}{2} & \frac{4}{2} & \frac{5}{2} & \cdots \\[4pt]
\frac{1}{3} & \frac{2}{3} & \frac{3}{3} & \frac{4}{3} & \frac{5}{3} & \cdots \\[4pt]
\frac{1}{4} & \frac{2}{4} & \frac{3}{4} & \frac{4}{4} & \frac{5}{4} & \cdots \\[4pt]
\frac{1}{5} & \frac{2}{5} & \frac{3}{5} & \frac{4}{5} & \frac{5}{5} & \cdots \\[4pt]
\cdot & \cdot & \cdot & \cdot & \cdot & \cdots \\
\cdot & \cdot & \cdot & \cdot & \cdot & \cdots
\end{array}
$$

Draw a zig-zag line through this array as follows:

$$
\begin{array}{ccccccc}
\frac{1}{1} \rightarrow & \frac{2}{1} & \frac{3}{1} \rightarrow & \frac{4}{1} & \frac{5}{1} & \cdots \\[4pt]
\frac{1}{2} & \frac{2}{2} & \frac{3}{2} & \frac{4}{2} & \frac{5}{2} & \cdots \\[4pt]
\frac{1}{3} & \frac{2}{3} & \frac{3}{3} & \frac{4}{3} & \frac{5}{3} & \cdots \\[4pt]
\frac{1}{4} & \frac{2}{4} & \frac{3}{4} & \frac{4}{4} & \frac{5}{4} & \cdots \\[4pt]
\frac{1}{5} & \frac{2}{5} & \frac{3}{5} & \frac{4}{5} & \frac{5}{5} & \cdots \\[4pt]
\cdot & \cdot & \cdot & \cdot & \cdot & \cdots
\end{array}
$$

We can now list the positive rationals by simply moving along the zig-zag line in the direction of the arrows, writing down each number as we reach it (and omitting numbers we have already written down, such as $\frac{2}{2}, \frac{3}{3}, \frac{4}{2}$ and so on). The list starts like this:

$$
1, \, 2, \, \frac{1}{2}, \, \frac{1}{3}, \, 3, \, 4, \, \frac{3}{2}, \, \frac{2}{3}, \, \frac{1}{4}, \, \frac{1}{5}, \, 5, \, \ldots .
$$

Thus we obtain a complete list of all the positive rationals, showing that $\mathbb{Q}^+$ is countable.

Finally, we need to deduce that $\mathbb{Q}$ is countable. Let the above list of the elements of $\mathbb{Q}^+$ be $\mathbb{Q}^+ = \{q_1, q_2, q_3, \ldots\}$. Then we can list the elements of $\mathbb{Q}$ as

$$
\mathbb{Q} = \{0, q_1, -q_1, q_2, -q_2, q_3, -q_3, \ldots\} \, ,
$$

which shows that $\mathbb{Q}$ is countable. ∎

The next proposition provides a quite useful method for showing that sets are countable.

PROPOSITION 19.4

Let S be an infinite set. If there is a 1-1 function $f : S \to \mathbb{N}$, then S is countable.

PROOF   Recall that the image of f is the set

$$f(S) = \{f(s) \mid s \in S\} \subseteq \mathbb{N}.$$

Since f is 1-1, $f(S)$ is an infinite set. Therefore by Proposition 19.2, $f(S)$ is countable. Consequently there is a bijection $g : \mathbb{N} \to f(S)$.

Now we can regard f as a function from S to $f(S)$. As such, f is onto; hence as f is also 1-1, f is a bijection from S to $f(S)$. There is therefore an inverse function $f^{-1} : f(S) \to S$.

Finally, consider the composition $f^{-1} \circ g : \mathbb{N} \to S$. By Proposition 18.2(c), this is a bijection. This means that S is countable.   ∎

This proposition can be used in many further examples:

Examples

(1) Here is another proof that $\mathbb{Q}^+$ is countable. Define $f : \mathbb{Q}^+ \to \mathbb{N}$ by

$$f(\frac{m}{n}) = 2^m 3^n$$

where $m, n \in \mathbb{N}$ and $\frac{m}{n}$ is in lowest terms. Then f is 1-1, since

$$f(\frac{m}{n}) = f(\frac{p}{q}) \implies 2^m 3^n = 2^p 3^q \implies m = p, n = q$$

using the Fundamental Theorem of Arithmetic 12.1. Hence $\mathbb{Q}^+$ is countable by Proposition 19.4.

(2) A very similar proof shows that the Cartesian product $\mathbb{N} \times \mathbb{N}$ is countable: just define $f : \mathbb{N} \times \mathbb{N} \to \mathbb{N}$ by $f(m, n) = 2^m 3^n$, and observe again that f is 1-1. Likewise, $\mathbb{N} \times \mathbb{N} \times \mathbb{N}$ is countable, since the function $g(m, n, l) = 2^m 3^n 5^l$ from $\mathbb{N} \times \mathbb{N} \times \mathbb{N} \to \mathbb{N}$ is 1-1; and so on — the Cartesian product of any finite number of copies of $\mathbb{N}$ is countable.

An Uncountable Set

We have shown that many sets are countable. Are there in fact any infinite sets which are *not* countable? The answer is yes. Here is the most famous example of an uncountable set. (An *uncountable* set is an infinite set which is not countable.)

THEOREM 19.1

The set $\mathbb{R}$ of all real numbers is uncountable.

PROOF   We prove the theorem by contradiction. So suppose that $\mathbb{R}$ is countable. This means that we can list all the elements of $\mathbb{R}$ as

$$\mathbb{R} = \{r_1, r_2, r_3, \ldots\} \ .$$

Express each of the r_is in the list as a decimal:

$$r_1 = m_1.a_{11}a_{12}a_{13}\ldots$$
$$r_2 = m_2.a_{21}a_{22}a_{23}\ldots$$
$$r_3 = m_3.a_{31}a_{32}a_{33}\ldots$$
$$\cdot \quad \cdot \quad \cdot$$
$$r_n = m_n.a_{n1}a_{n2}a_{n3}\ldots$$
$$\cdot \quad \cdot \quad \cdot$$

(where each $m_i \in \mathbb{Z}$ and each $a_{ij} \in \{0, 1, 2, \ldots, 9\}$).

Now define a real number $r = 0.b_1b_2b_3\ldots$ as follows.

To choose the first decimal digit b_1: if $a_{11} \neq 1$, let $b_1 = 1$; and if $a_{11} = 1$, let $b_1 = 2$. (Hence $b_1 \neq a_{11}$.)

To choose the second decimal digit b_2: if $a_{22} \neq 1$, let $b_2 = 1$; and if $a_{22} = 1$, let $b_2 = 2$. (Hence $b_2 \neq a_{22}$.)

And so on: in general, to choose the nth decimal digit b_n: if $a_{nn} \neq 1$, let $b_n = 1$; and if $a_{nn} = 1$, let $b_n = 2$. (Hence $b_n \neq a_{nn}$.)

In this way we define a real number $r = 0.b_1b_2b_3\ldots$. Since $r_1, r_2, r_3, \ldots$ is a list of *all* real numbers, r must belong to this list, so $r = r_n$ for some n. But $b_n \neq a_{nn}$, so r and r_n differ in their n^{th} decimal digit. Note also that r does not end in recurring 9s or 0s (all the b_is are 1 or 2). Hence $r \neq r_n$, which is a contradiction.

Therefore there is no bijection from $\mathbb{N}$ to $\mathbb{R}$, which is to say that $\mathbb{R}$ is un-countable.   ∎

This famous theorem is due to Georg Cantor (1874), the founder of modern set theory. The wonderfully clever idea of the proof — defining the decimal $r = 0.b_1b_2b_3\ldots$ by adjusting the "diagonally placed" decimal digits a_{nn} in the array of r_is — is called Cantor's "diagonal argument," and can be used to prove that all sorts of other sets are uncountable. (See Exercise 2 at the end of the chapter.)

A consequence of Theorem 19.1 is that the set of irrationals $\mathbb{R} - \mathbb{Q}$ is uncountable — for if it were countable, then $\mathbb{R}$ would be the union of two countable sets, and hence would be countable. Thus, there are in some sense "more" irrational numbers than there are rationals.

A Hierarchy of Infinities

DEFINITION   *Let A and B be sets. If A and B are equivalent to each other, i.e., there is a bijection from A to B, we say that A and B have the same* cardinality, *and write* $|A| = |B|$.

If there is a 1-1 function from A to B we write $|A| \leq |B|$.

And if there is a 1-1 function from A to B, but no *bijection from A to B we write* $|A| < |B|$, *and say that A has smaller cardinality than B. (Thus,* $|A| < |B|$ *is the same as saying that* $|A| \leq |B|$ *and* $|A| \neq |B|$.)

According to this definition, we have

$$|\mathbb{N}| = |\mathbb{Q}| = |\mathbb{N} \times \mathbb{N}| \, ,$$

and

$$|\mathbb{N}| < |\mathbb{R}| \, .$$

Thus there are at least two different types of "infinity," namely $|\mathbb{N}|$ and $|\mathbb{R}|$.

Are there more types of infinity? For example, is there a set of greater cardinality than $\mathbb{R}$?

The answer is yes, and again this is due to Cantor. To understand this, we first need a definition.

DEFINITION   *If S is a set, let P(S) be the set consisting of all the subsets of S.*

For example, if $S = \{1, 2\}$ then $P(S) = \{\{1, 2\}, \{1\}, \{2\}, \emptyset\}$. In general, if S is a finite set of size n, then $|P(S)| = 2^n$ (Exercise 5 at the end of Chapter 16).

Cantor's theory is based on the following result.

PROPOSITION 19.5

Let S be a set. Then there is no bijection from S to P(S). Consequently, $|S| < |P(S)|$.

Using the proposition we obtain a hierarchy of infinities, starting at $|\mathbb{N}|$:

$$|\mathbb{N}| < |P(\mathbb{N})| < |P(P(\mathbb{N}))| < |P(P(P(\mathbb{N})))| < \ldots$$

Thus there are indeed many types of "infinity."

Proof of Proposition 19.5   This is a very subtle proof by contradiction. You may well have to go through it a few times before really understanding it.

Suppose there is a bijection $f : S \to P(S)$. Then *every* subset of S is equal to $f(s)$ for some $s \in S$.

For any $s \in S$, $f(s)$ is a subset of S, and it is certainly the case that either $s \in f(s)$ or $s \notin f(s)$. (For example there exists s_1 such that $f(s_1) = S$, and then $s_1 \in f(s_1)$; likewise, there exists s_2 such that $f(s_2) = \emptyset$, and then $s_2 \notin f(s_2)$.) Define A to be the set of all elements s of S such that $s \notin f(s)$; symbolically,

$$A = \{s \in S \mid s \notin f(s)\} .$$

(In the above notation, $s_1 \notin A$ but $s_2 \in A$.)

Certainly A is a subset of S, that is, $A \in P(S)$. Therefore, as f is a bijection, $A = f(a)$ for some $a \in S$.

We now ask the question: does a belong to A?

If $a \notin A$, then $a \notin f(a)$, so by definition of A, we have $a \in A$. This is a contradiction. And if $a \in A$ then $a \in f(a)$, so by definition of A we have $a \notin A$, again a contradiction.

Thus we have reached a contradiction in any case. So we conclude that there cannot be a bijection from S to $P(S)$.

Since there is certainly a 1-1 function $f : S \to P(S)$, namely $f(s) = \{s\}$ for all $s \in S$, it follows that $|S| < |P(S)|$, and the proof is complete.   ∎

I hope you like this argument. If you do, you are undoubtedly a budding pure mathematician. (And even if you do not, you may well find lots of other branches of pure mathematics to tickle your fancy.) Good luck in your future studies.

It is hard to think of a better place to stop, so I will stop.

Exercises for Chapter 19

1. (a) Show that if A is a countable set and B is a finite set, then $A \cup B$ is countable.

   (b) Show that if A and B are both countable sets, then $A \cup B$ is countable.

2. Let S be the set consisting of all infinite sequences of 0s and 1s (so a typical member of S is 010011011100110....., going on forever). Use Cantor's diagonal argument to prove that S is uncountable.

3. Let S be the set consisting of all the finite subsets of $\mathbb{N}$. Prove that S is countable.

4. Every night, critic Ivor Smallbrain gets drunk, staggers out of the pub, and performs a kind of random walk towards his home. At each step of this walk, he stumbles either forwards or backwards, and the walk ends either when he collapses in a heap or when he reaches his front door (one of these always happens after a finite [possibly very large] number of steps). Ivor's Irish friend Gerry O'Laughing always accompanies him, and records each random walk as a sequence of 0s and 1s: at each step he writes a 1 if the step is forwards and a 0 if it is backwards.

   Prove that the set of all possible random walks is countable.

Further Reading

More on the real numbers and analysis:

1. K. Binmore, *Mathematical Analysis: A Straightforward Approach,* Cambridge University Press, 1982.
2. R. Haggarty, *Fundamentals of Mathematical Analysis,* Addison-Wesley, 1993.

More on the complex numbers:

3. M.R. Spiegel, *Complex Variables,* Schaum's Outline Series, McGraw-Hill, New York, 1974.

More on proofs, induction and Euler's Formula:

4. I. Lakatos, *Proofs and Refutations,* Cambridge University Press, 1976.
5. G. Polya, *Induction and Analogy in Mathematics,* Princeton University Press, Princeton, NJ, 1954.
6. D.J. Velleman, *How to Prove it: A Structured Approach,* Cambridge University Press, 1994.

More on integers, sets and functions:

7. P. Eccles, *An Introduction to Mathematical Reasoning,* Cambridge University Press, 1997.

More on Number Theory:

8. K.H. Rosen, *Elementary Number Theory and its Applications,* Addison-Wesley, 1993.

History:

9. D.M. Burton, *The History of Mathematics: An Introduction,* McGraw-Hill, New York, 1997.

General:

10. R. Courant and H. Robbins, *What is Mathematics?,* Oxford University Press, 1979.

Index of Symbols

$\emptyset$	empty set, 2
$\in$	belongs to, 2
$\notin$	does not belong to, 2
$\subseteq$	contained in, 2
$\Rightarrow$	implies, 3
$\Leftrightarrow$	if and only if, 3
$\mathbb{R}$	the set of real numbers, 11
$\mathbb{Z}$	the set of integers, 11
$\mathbb{N}$	the set of natural numbers, 11
$\mathbb{Q}$	the set of rational numbers, 12
$x^{m/n}$	rational power, 29
i	square root of -1, 33
$Re(z)$	real part of z, 33
$Im(z)$	imaginary part of z, 33
$\mathbb{C}$	the set of complex numbers, 33
$|z|$	modulus of z, 34
$arg(z)$	argument of z, 35
$re^{i\theta}$	polar form, 39
$n!$	n factorial, 58
LUB	least upper bound, 80
GLB	greatest lower bound, 80
$a|b$	a divides b, 87
$\text{hcf}(a, b)$	highest common factor, 88
$a \equiv b \bmod m$	a congruent to b modulo m, 107
$\binom{n}{r}$	binomial coefficient, 117
$\binom{n}{r_1, \ldots, r_k}$	multinomial coefficient, 120
$A \cup B$	union of sets A and B, 127
$A \cap B$	intersection of A and B, 127
$A - B$	difference of sets A and B, 129
$f : S \rightarrow T$	function from S to T, 137
$f(S)$	image of function f, 137
ι_S	identity function on S, 138
f^{-1}	inverse function, 140
$g \circ f$	composition of functions, 140
$P(S)$	set of all subsets of S, 151

Index